D0437997

# REBUILDING URBAN NEIGHBORHOODS

# Cities & Planning Series

The *Cities & Planning Series* is designed to provide essential information and skills to students and practitioners involved in planning and public policy. We hope the series will encourage dialogue among professionals and academics on key urban planning and policy issues. Topics to be explored in the series may include growth management, economic development, housing, budgeting and finance for planners, environmental planning, GIS, small-town planning, community development, and community design.

## Series Editors

Roger W. Caves, Graduate City Planning Program,
   *San Diego State University*

Robert J. Waste, Department of Political Science,
   *California State University at Chico*

Margaret Wilder, Department of Geography and Planning,
   *State University of New York at Albany*

## Advisory Board of Editors

Edward J. Blakely, *University of Southern California*
Robin Boyle, *Wayne State University*
Linda Dalton, *California Polytechnic State University, San Luis Obispo*
George Galster, *Wayne State University*
Eugene Grigsby, *University of California, Los Angeles*
W. Dennis Keating, *Cleveland State University*
Norman Krumholz, *Cleveland State University*
John Landis, *University of California, Berkeley*
Gary Pivo, *University of Washington*
Daniel Rich, *University of Delaware*
Catherine Ross, *Georgia Institute of Technology*

W. Dennis Keating
Norman Krumholz
*Editors*

# REBUILDING URBAN NEIGHBORHOODS

## Achievements, Opportunities, and Limits

Cities & Planning

**SAGE Publications**
*International Educational and Professional Publisher*
Thousand Oaks   London   New Delhi

*For information:*

Sage Publications, Inc.
2455 Teller Road
Thousand Oaks, California 91320
E-mail: order@sagepub.com

Sage Publications Ltd.
6 Bonhill Street
London EC2A 4PU
United Kingdom

Sage Publications India Pvt. Ltd.
M-32 Market
Greater Kailash I
New Delhi 110 048 India

Printed in the United States of America

*Library of Congress Cataloging-in-Publication Data*

Main entry under title:

Rebuilding urban neighborhoods: Achievements, opportunities,
 and limits / edited by W. Dennis Keating,
Norman Krumholz.
   p. cm. — (Cities and planning; v. 5)
  Includes bibliograhical references and index.
  ISBN 0-7619-0691-6 (cloth: alk. paper)
  ISBN 0-7619-0692-4 (pbk.: alk. paper)
  1. Urban renewal—United States.  I. Keating, W. Dennis (William
Dennis)  II. Krumholz, Norman.  III. Series: Cities and planning
series; v. 5.
  HT175.R425  1999
  304.3′416′0973—dc21                                    98-40295


99  00  01  02  03  10  9  8  7  6  5  4  3  2  1

| | |
|---|---|
| *Acquiring Editor:* | Catherine Rossbach |
| *Editorial Assistant:* | Heidi Van Middlesworth |
| *Production Editor:* | Diana E. Axelsen |
| *Editorial Assistant:* | Nevair Kabakian |
| *Typesetter/Designer:* | Janelle LeMaster |
| *Indexer:* | Mary Mortensen |

# CONTENTS

# SERIES EDITORS' INTRODUCTION

The study of cities is a dynamic, multifaceted area of inquiry that combines a number of disciplines, perspectives, and time periods, as well as numerous actors. Urbanists alternate between examining one issue through the eyes of a single discipline and looking at the same issue through the lenses of a number of disciplines to arrive at a holistic view of cities and urban issues. The books in this series look at cities from a multidisciplinary perspective, affording students and practitioners a better understanding of the multiplicity of issues facing planning and cities, and of emerging policies and techniques aimed at addressing those issues. The series focuses on traditional planning topics such as economic development, management and control of growth, and geographic information systems but also includes broader treatments of conceptual issues embedded in urban policy and planning theory.

The impetus for the Cities & Planning Series originates in our reaction to a common recurring event—the ritual selection of course textbooks. Although we all routinely select textbooks for our classes, many of us are never completely satisfied with the offerings. Our dissatisfaction stems from the fact that most books are written for either an academic or practitioner audience. Moreover, on occasion, it appears as if this gap continues to widen. We wanted to develop a multidisciplinary series of manuscripts that would bridge the gap between academia and professional practice. The books are designed to provide valuable information to students/instructors and to practitioners by going beyond the narrow confines of traditional disciplinary boundaries to offer new insights into the urban field.

Dennis Keating and Norman Krumholz lead a distinguished group of authors in the timely and provocative text *Rebuilding Urban Neighborhoods*. They succeed in avoiding the usual pitfalls of edited texts by providing a coherent framework for understanding the limitations of urban policy, and by presenting a diverse yet complementary set of case studies that reveal the reality of community revitalization efforts to date. The case studies are rich, original discussions of local experiences in community rebuilding. Whether exploring the contradictions of efforts to revive South Central Los Angeles or negotiating the politics of Atlanta's Olympic Stadium project, the authors succeed in extending our understanding of the common threats and potential of urban communities to rebuild their social and economic foundations. The cases provide a portrait of the distinct character of individual communities, but they reveal even more vividly the commonality of human struggle and resilience in the face of daunting challenges. The book takes us on a journey through some of America's most important policy terrain and challenges us to consider new pathways of thought and action.

—Roger W. Caves
*San Diego State University*

—Robert J. Waste
*California State University at Chico*

—Margaret Wilder
*University of Delaware*

# PREFACE

The late 20th century in the United States can be regarded as the best and worst of times. The U.S. economy continues a historically protracted period of prosperity and growth, the Cold War is over and the hopes for peace are more optimistic and realistic, and a new millennium beckons. Yet, many whose jobs have been eliminated or who are not prepared for work in the information age are losing ground. The real wages of most American workers have been stagnant over the past two decades, while the wealthy have grown more so. The gap between rich and poor has been widening. Despite advances in racial understanding, the United States remains a racially divided society in all too many ways. Racial preference policies have become a political lightning rod. An attempted national dialogue on race in 1998 failed to take off. Many American cities have redeveloped their downtowns, reshaping their skylines, riverfronts, and harbors at enormous public expense, yet they have neglected neighborhoods and concentrations of poverty that have grown. The gulf between affluent newer suburbs and declining central cities remains a continuing phenomenon across most metropolitan areas.

This book seeks to understand the prospects for more successfully addressing these persistent urban problems in some of the most distressed and poorest neighborhoods in American cities. We look at the history of these neighborhoods and past efforts, mostly funded by the federal government, to solve these problems. In a conservative era of distrust of government when few national political leaders address urban problems, and when federal budget balancing, deregulation, and devolution of responsibility from Washington, D.C., to states and local government are all the vogue, it is difficult to see how such serious issues can be resolved. Almost certainly, there will be fewer public resources available to address persistent poverty in inner-city neighborhoods.

As we and the contributors recount, however, there has been some progress in many of these neighborhoods. Even in devastated areas like the South Bronx in New York City, there are promising signs of revitalization. The key to successful revitalization efforts has been the participation of residents and of community organizations and institutions. Most prominent have been community development corporations (CDCs) and churches. These organizations have struggled against these discouraging urban trends and what seems to be indifference on the part of government, media, and those who do not reside in central cities. In the face of heavy odds, they have achieved successes.

This book highlights some examples of these achievements. The contributors are also realistic in their appraisal of the difficulties facing the communities profiled and their residents. It is our hope that there will once again be serious attention paid to the still urgent urban crisis.

# ACKNOWLEDGMENTS

We would like to acknowledge the excellent editorial guidance provided by Professor Margaret Wilder (University of Delaware), one of the editors of the Cities & Planning Series. We also thank Catherine Rossbach and the Sage Publications staff for their support. Harold Jackson of the Cleveland State University College of Law staff produced the manuscript.

# CHAPTER 1

# INTRODUCTION

## W. Dennis Keating

On March 4, 1908, a devastating fire raced through Cleveland's Lakeview Elementary School, killing 174 people—all but 2 of them children. It was one of the worst school disasters in American history. The whole Cleveland community mourned. One account of the aftermath stated that

> lavishly dressed women, in expensive clothes and other objects of wealth, consoled women with humble shawls over their heads. Many foreign-born women, wearing well-worn dresses that showed they were on the brink of poverty . . . cradled and held the rich and well-to-do of the village in their arms, as the entire community's grief spanned all social status. (*Cleveland News*, March 6, 1908, p. 1)[1]

The same might have been written about any similar tragedy occurring at the same time in any American industrial city when rich and poor lived closely together. But today, the consequences of a similar tragedy would be much more narrowly felt, economically speaking. In 1997, of the children in the elementary school closest to the site of the old Lakeview school, 72% are African American and 83% come from households living under the poverty line. The rich and poor who used to be able to share in each other's lives now inhabit different worlds, and many of the poor families who remain in America's inner cities are facing increasing difficulties.

This is a book about the most distressed neighborhoods in American cities, about key events in their recent history, about federal policies that have affected their fate, and about the individuals and organizations that are helping shape their present and future.

Although the focus of this book is on distressed urban neighborhoods, concentrations of poverty in certain central city neighborhoods also have regional and national significance. These neighborhoods contain a relatively small percentage of the total U.S. population but a large percentage of poverty, crime and violence, joblessness, and children with bad public education. These factors undermine both our nation's social cohesion and its economic efficiency (Downs, 1994; Goldsmith & Blakely, 1992; Rusk, 1993). Without better schooling in our inner-city schools that will enable their students to find jobs, it is hard to see how our nation can retain its international competitiveness or economic well-being in the future (Kozol, 1991). Nor is it hard to see how rising barriers of race, ethnicity, and income inevitably will erode the social fabric of American society and our nation's social peace.

Resolution of the problems of these distressed neighborhoods remains an urgent need. As the case studies of such neighborhoods in this book demonstrate, these problems are persistent. The examples and data on urban poverty suggest that the number of distressed urban neighborhoods and their residents living in poverty have increased significantly since 1970. Despite an improved economic outlook in the 1990s, they are not enjoying the benefits of prosperity.

## THE DECLINE OF
## URBAN NEIGHBORHOODS

Distressed neighborhoods are those that simultaneously exhibit disproportionately high levels of poverty, joblessness, female-headed households, and dependency on welfare assistance. Most of America's severely distressed neighborhoods are found in our 100 largest cities, and their number is growing. In 1990, 11% of the population of the nation's 100 largest cities lived in extreme poverty neighborhoods (with 40% or more of the residents living in poverty), compared to 8% in 1980 and 5% in 1970 (Kasarda, 1993). Between 1970 and 1990, the actual number of people living in concentrated poverty neighborhoods grew from 3.8 million to 10.4 million (Mincy & Weiner, 1993). High poverty rates in these neighborhoods vary by race, ethnicity, and region, with the greatest concentrations found in predominantly minority areas of the older central cities (Jargowsky, 1997).

After mid-century, many distressed urban neighborhoods and their cities had to face the massive problems of industrial decline, a shrinking tax base, an automobile culture fueling urban sprawl, a breakdown in family structure, racial tensions, crime, and drugs. Any one of these disasters would have been enough to shake an urban area, but it was the particular misfortune of the most distressed of these urban neighborhoods to experience all of them. As early as the mid-1960s, many of the neighborhoods described in this book had embarked on a pattern of seemingly irreversible long-term decline.

As people and jobs left, buildings were abandoned, and devastation spread, the media reached into the past to describe the stricken areas. In the South Bronx, the ground zero of perhaps the most devastated neighborhood in the United States, the police department's 41st Precinct headquarters was described in a 1981 film as *Fort Apache, the Bronx*. The reputation of the neighborhood as a lawless wasteland was further exacerbated in Tom Wolfe's 1987 novel *The Bonfire of the Vanities*. Despite the devastation, as current revitalization efforts have revealed, there remain among the ruins survivors who have not given up. They believe, in a testimony to the durability and resiliency of the human spirit, that their environmental plight is not beyond redemption.

The devastation of these neighborhoods and the impact on their residents and the cities in which they are located is vividly portrayed by sociologist and photographer Camilo Vergara's *The New American Ghetto* (1995). Vergara's camera shoots scenes from ghetto neighborhoods in Camden, Chicago, Cleveland, Detroit, Los Angeles, Miami, and New York, all cities featured in this book. He also concentrates on Gary and Newark. Although Vergara's outlook is generally bleak and pessimistic, he does acknowledge ongoing efforts to save and rebuild many of these areas. Critical to these efforts is the role of the federal government.

## FEDERAL INTERVENTION

The social and physical problems of these distressed neighborhoods and their people have long been the target of governmental and philanthropic programs. Indeed, it might be said that since the 1950s, neighborhood initiatives have been shaped by federal programs and the responses of local governments to them. The 1949 Federal Housing Act set the stage for a decade of programs that aggravated changing urban economic and social patterns, including the movement of African Americans to the cities. Urban renewal and slum clearance led to the demolition of large numbers of affordable housing units that were never replaced (Halpern, 1995). The federal highway system, in its search for lowest cost routes in the cities, uprooted many urban minority communities and forced their relocation within the cities in segregated housing markets. By permitting (or actually encouraging) racial discrimination until the 1970s in the Federal Housing Administration's mortgage underwriting program and in various housing subsidy programs, federal policies facilitated white middle-class movement to the suburbs and trapped African Americans in older city neighborhoods. In the process, most African Americans, unable to buy suburban homes, were cut off from the powerful middle-class wealth-generating machine of appreciating housing value.

Urban racial and economic segregation was further reinforced by the location and design of public housing projects, many of which were built to house the displacees of urban renewal and highway projects. By the 1990s, many of these same public housing projects were the locus of extreme economic and social distress. This was poignantly personalized in the portrayal of the life of two young African American residents of Chicago's Henry Horner Homes in *There Are No Children Here: The Story of Two Boys Growing Up in the Other America* (Kotlowitz, 1991). The destruction of the abandoned high-rise Pruitt-Igoe project in St. Louis in 1973 seemed to signify the eventual demise of this kind of isolated, ghettoized project elsewhere in the United States.

The 1960s and 1970s saw the federal government's interest in neighborhood initiatives at an unprecedented level. Community health and service centers were supported. The Concentrated Code Enforcement Program was designed to improve housing and stabilize neighborhoods without the disruptive demolition of urban renewal; the Community Action Program attempted to forge a coherent attack on poverty while ensuring that neighborhood residents had a voice in the planning; and the Model Cities program tried to achieve the same ends without stirring up a political backlash. In the Carter administration, the Neighborhood Self-Help Development Program aimed to stimulate grassroots initiatives, and the Urban Development Action Grant Program was to provide necessary project financing to stimulate distressed central city employment. All struggled with internal contradictions: Should decision making be vested in City Hall or in the people of the community? Should the goal of neighborhood revitalization be racial integration or gilding the ghetto? Should activists confront or negotiate with banks and other powerful mainstream institutions adversely affecting the community through such policies as redlining?

## FEDERAL CUTBACKS

In the 1980s and 1990s, Presidents Reagan and Bush—proponents of the New Federalism proclaimed by President Nixon—proved that federal neglect and underfunding could be as disastrous to troubled urban neighborhoods as misguided earlier programs. They cut direct spending on cities sharply. In 1980, federal contributions made up 18% of city budgets; by 1990, federal contributions had dropped to 6.4% (Wilson, 1996). The result was a cutback in basic services at a time when cities were further challenged by rising tides of poverty, the AIDS epidemic, and a sharp rise in the homeless population.

The first-term Clinton administration was rhetorically activist but fiscally strapped. It also took centrist positions on many issues. In the administration of Housing and Urban Development (HUD) Secretary Henry Cisneros, HUD struggled with questions of how best to help troubled urban neighborhoods: by trying to revitalize them as places, or providing the residents with skills, transportation, links, and access to suburban jobs (Gottlieb, 1997)? Ultimately, the Clinton administration attempted both neighborhood and regional approaches, with the Empowerment Zone program attempting to revitalize distressed central city neighborhoods and the Moving to Opportunity Program providing poor central city residents with housing opportunities in the surrounding suburbs.

Given these contradictory strains in public policy and the American preference for trying to accomplish a lot with a little in the area of social reform, it is not surprising that for all these long-standing efforts—from the public housing program of the 1930s to the Empowerment Zone program enacted under President Bill Clinton in 1994 (a belated federal response to Los Angeles's 1992 South Central riot)—the poverty and social problems of distressed urban neighborhoods in many U.S. cities persist and have deepened. Meanwhile, many Americans have reacted to concentrated poverty and associated social problems in the central cities by distancing themselves from these issues through continued suburbanization, gated communities, and living very private lives. The flow of population and jobs to the suburbs accelerated between 1980 and 1990. In 1950, central cities contained 57% of metropolitan area residents and 70% of metro area jobs. By 1990, central cities contained only 37% of all metro area residents and 45% of all jobs (Mieszkowski & Mills, 1993). Between 1980 and 1990, the suburbs captured most of the net job growth in manufacturing employment, while central cities consistently lost manufacturing employment (Hughes & Sternberg, 1992). In sum, population, private investment, jobs, and economic opportunity all have been distancing themselves from distressed central city neighborhoods and their problems, and the rate of outmigration into surrounding regions continues (Keating, Krumholz, & Star, 1996). This means that the inner-city poor, especially minorities, face daunting challenges in attaining even a modest level of economic well-being. The futures of these individuals and the neighborhoods they inhabit are in serious jeopardy.

This book reviews a wide range of national and local policies and programs that have been aimed at the most distressed neighborhoods in U.S. cities. Selected case studies reveal the positive and negative impacts of federal and local policies and programs; key events in the history of these neighborhoods; and key organizers, community organizations, politicians, and institutions affecting these neighborhoods. The book concludes with speculations on the future of these distressed neighborhoods and recommendations on policies and programs that might be required to successfully eradicate poverty, racism, and inequality and restore these neighborhoods and their people.

## CITIES AND DISTRESSED NEIGHBORHOODS: CASE STUDIES

### Atlanta (Peoplestown)

Peoplestown is a small neighborhood located south of downtown Atlanta. It consists of only 2,527 persons, 50% of whom are poor and 95% of whom are African American. Peoplestown is one of four poor neighborhoods selected to split $65 million of Empowerment Zone and Community Development Block Grant (CDBG) Section 108 funds. Since the 1960s, Peoplestown has suffered from highway construction, urban renewal, and sports-related development.

The chapter by Larry Keating describes the Peoplestown Revitalization Corporation (PRC), which is leading indigenous neighborhood redevelopment efforts aimed at stabilizing the community. Keating recounts Peoplestown's struggle to share in the benefits of the 1996 Atlantic Olympics. The ultimate fate of this neighborhood remains problematic.

### Camden

In his chapter, Robert Catlin explores the problems and opportunities of Camden, New Jersey. His analysis and conclusions end on a somber note. Camden is an extremely troubled part of urban America in the 1990s. It is an older industrial city that has been in decline since 1950. Its present population is about half African American and one quarter Hispanic, and very poor. The city has been losing jobs, housing, and economic investment of all sorts for decades. Still, certain opportunities remain: The city is close to central Philadelphia; it has large, easily assembled industrial sites; it has excellent transportation links; and it has a supportive state government that, by 1990, was providing one quarter of the city's budget. For a number of reasons, however, Camden has been unable to capitalize on its potential for redevelopment. Catlin offers four controversial recommendations for Camden's revitalization that address issues of mayoral leadership, land use planning, government consolidation, and neighborhood-based development.

Although Camden was named an Empowerment Zone (EZ) jointly with Philadelphia by President Clinton in 1994, current efforts in the city's EZ have not yet been effective. In Catlin's view, the present EZ can assist in the process of change but cannot substitute for the four important interventions suggested above.

## Chicago (North Lawndale)

In his chapter on the North Lawndale neighborhood in Chicago, Robert Giloth describes a 60-year process during which the neighborhood went from a white, working-class ethnic model of stability and cohesion in the 1940s to an emblem of failed Great Society policies and an example of the problems of African Americans suffering from the effects of concentrated poverty and racial segregation in the 1970s, and then to its present situation in the 1990s as an area attracting significant new investment and, perhaps, beginning an upward turnaround.

Three innovations occurred in North Lawndale, spurring the most recent change: Activities of the Lawndale Christian Development Corporation (LCDC); Homan Square; and the investments of the Steans Foundation. These efforts were stimulated further by the growing Illinois Medical Center and the new United Center sports stadium to the east of the area and the growing Mexican American community of Little Village and its high retail sales volume to the south.

Giloth's North Lawndale story is about the diaspora of African Americans to northern cities and about what they found there. The history of community development in this neighborhood during the last 60 years offers an array of organizing, planning, and development efforts, many of which proved frustrating and unsuccessful. The latest generation of community building, however, may succeed in producing a neighborhood future that is more promising than ever.

## Cleveland (Hough and Central)

Norman Krumholz's chapter on Hough and Central describes racial segregation and deterioration at the core of the city, Cleveland's ambitious urban renewal and public housing programs of the 1950s and 1960s, rapid white flight and racial change in the 1960s and 1970s accompanied by two racial riots, and the gradual sowing of the seeds of a limited neighborhood revival in the 1990s.

The Central neighborhood was and is the poorest neighborhood in the city, with about 6,000 units of subsidized housing, most of it owned and operated by the Cuyahoga County Metropolitan Housing Authority (CMHA). CMHA received about $100 million above its normal budget in the early 1990s and has put about 2,400 units into substantial rehabilitation while "thinning out" its remaining stock through demolition. New single units at market rates are also beginning to fill the vacant land of the Central neighborhood. Those concerned with housing the poor, however, worry about the prospects of providing decent shelter for low-income families with fewer units of affordable housing.

To an extent, the same pattern of new housing construction on vacant land is being followed in the Hough neighborhood, located between Cleveland's two most important job locations. Although the neighborhood continues to be very poor, new single-family houses are being built here at prices ranging from $65,000 to $645,000. Meanwhile, Hough had a 1990 median housing sale price of $21,600.

Hough is one of Cleveland's three Supplementary Empowerment Zone (SEZ) neighborhoods. Krumholz closes this chapter by speculating on the probability that the SEZ will be able to restore these neighborhoods to social and economic viability.

## Detroit

Detroit represents a half century of redevelopment efforts, which Mittie O. Chandler analyzes in detail. Detroit, the site of major riots in 1943 and 1967, presents one of the most important examples of urban decline and problems of distressed neighborhoods. The causes of the city's long-standing loss of population and jobs, building abandonment, and racial strife are explored, as are efforts to revive its inner city.

Chandler reviews the inability of such federal programs as urban renewal, model cities, and the Community Development Block Grant program to redevelop both Detroit's central business district and its neighborhoods. Now, the question is whether Detroit, under a new mayor, Dennis Archer, can begin to reverse past setbacks, attract private reinvestment in its Empowerment Zone, and sustain community development. Chandler recounts the development of the city's application to HUD, including conflicts between City Hall and community organizations over representation in the process, and ultimately the establishment of a new, autonomous entity to oversee implementation of the zone's activities.

Although the eventual outcomes are too difficult to predict in the early stages, evidence of optimism exists with commitments from the corporate and financial sectors to the empowerment zone effort. Given the magnitude and persistence of economic declines, however, whether the empowerment zone designation will make a discernible difference for the city and its residents remains to be seen.

## East St. Louis (Winstanley)

Whereas Atlanta's Peoplestown is a small, poor neighborhood in a relatively prosperous central city, the Winstanley/Industrial Park neighborhood in East St. Louis is one of the poorest neighborhoods in a city that has been described as the most distressed small city in America. East St. Louis, which is 98% African American and has 62% female-headed households and 39% of its residents in poverty, has multiple problems that have been thoroughly described on radio and television programs across the country, including *The Phil Donahue Show* and *60 Minutes*. Kenneth Reardon's chapter on Winstanley nevertheless offers an inspirational model of accomplishments through empowerment and grassroots leadership.

The Winstanley/Industrial Park revitalization effort is led by a local minister, with extensive planning assistance from faculty and students of the University of Illinois at Urbana-Champaign. Together, they have developed an empowerment model of neighborhood planning that involves extensive citizen participation and transforms the residents of a community from passive objects to active subjects.

The group established neighborhood goals and created a nonprofit community development corporation, Winstanley/Industry Park Neighborhood Organization (WIPNO), to implement the community's future revitalization efforts. Initially, WIPNO focused on small-scale, self-help projects, but Reardon recounts how over a 6-year period WIPNO developed a children's playground, established the East St. Louis Farmer's Market (now generating $395,000 in sales and $76,000 in employee wages for local residents), attracted charitable contributions as well as HOME and CDBG funds, rehabilitated a number of homes for low-income families, and built its organizational capacity. WIPNO became an important partner for the East St. Louis CDBG agency as well as an inspiration to other depressed East St. Louis neighborhoods to launch similar efforts.

The achievements of WIPNO so far are modest, compared to multimillion-dollar "big bang" downtown projects, but given the scope of East St. Louis's devastation, they are remarkable. Students and faculty are involved in a rich, applied learning effort, and the city, for the first time in its history, is pursuing an ambitious neighborhood revitalization program, in partnership with an expanding network of recently formed community-based development organizations.

## Los Angeles

Los Angeles was the scene of the Watts riot of 1965, which triggered several summers of urban riots in the United States. The 1992 South Central riot following the verdict in the Rodney King police brutality trial was yet another major urban disturbance in Los Angeles. This led to the creation of the federal Empowerment Zone program in 1994. Ironically, Los Angeles was not initially designated as one of the urban empowerment zones but was then awarded a Supplementary Empowerment Zone (SEZ) grant by HUD.

Ali Modarres focuses on the demography and geography of the Los Angeles supplementary empowerment zone. He analyzes the problems associated with attempting to deal with poverty in a city with a very mobile and changing population, including numerous immigrants. Latinos, not African Americans, make up the majority of the SEZ in Los Angeles. In reviewing the Los Angeles SEZ strategy, Modarres points out the pitfalls of a top-down redevelopment strategy dependent largely on a new Los Angeles Community Development Bank and the hope that business loans will spark new economic development and, therefore, the hiring of poor and unemployed SEZ residents.

## Miami (Overtown)

Miami's Overtown neighborhood is a picture of severe distress. From a densely populated neighborhood of 40,000 people in 1950, Overtown in 1990 had fallen to only 12,000 residents, 83% of whom were African American. The massive interchange of I-95, I-395, and State Road 836 has cut the neighborhood in three pieces and disrupted the street grid. Concentrated poverty characterizes Overtown, as does racial isolation. In 1990, fully 51% of all Overtown households earned less than $10,000 a year. Overtown has been the site of several race riots in recent years (Gale, 1996).

In his chapter on the Overtown neighborhood, Dennis Gale identifies four models employed over the last 40 years to assist the community: public housing and urban renewal, indigenous community development, historic restoration and cultural pride, and megastructures (an arena and a Metrorail station) and economic spillover. A fifth model of urban redevelopment was represented by the designation in 1994 of Miami-Dade County as a federal Empowerment Zone/Enterprise Community.

Redevelopment possibilities for Overtown seem to be bleak. Apparently, the most promising future for the neighborhood is to capitalize on its proximity to downtown and Biscayne Boulevard, on its transit linkage via its Metrorail station, and on a local hospital that is a large employer to build a multi-ethnic, multiracial neighborhood for singles and childless couples. It remains to be seen whether such a plan can overcome Overtown's proximity to major highways, existing subsidized housing developments, and sometimes bitter and contentious ethnic relationships.

## New York City (South Bronx and Red Hook)

A few of the neighborhoods in New York City discussed by Tom Angotti in his chapter conjure up stereotypical images of devastation and ruin. In the 1970s, it was high noon on the streets of the South Bronx, with widespread crime, abandonment, and arson prominent. More and more buildings burned, and then wrecking balls and bulldozers knocked down the hollow hulks.

Angotti tells us that things have changed in many New York neighborhoods. Significant portions of the South Bronx, for example, now have been rebuilt, with population levels stabilized and crime reduced almost everywhere. In many neighborhoods, there are signs of revitalization and a distinct sense of optimism. The condition of relative poverty, however, remains in the South Bronx and has worsened in other New York City neighborhoods such as Red Hook.

Angotti analyzes Red Hook, a relatively small, geographically isolated community on the South Brooklyn waterfront. Its population is about 50% African American and 42% Hispanic, and 70% of its families live in public housing. Although a plan for the regeneration of Red Hook was prepared by the community in 1994, its future is threatened by the current climate of privatization and deregulation at all levels of government.

At the local level, the weakening of local rent controls and regulations cushioning the processes of gentrification and displacement are worrisome. Federal contributions that found their way to distressed neighborhoods in the form of welfare, Medicaid, and food stamps are dwindling rapidly, while budget cuts at HUD threaten repairs and maintenance for low-income housing. So, while the sun is shining on some of the sidewalks of New York, other city streets are still in shadows.

The Empowerment Zone (EZ) initiative has yet to have any discernible impact on New York City because, according to Angotti, the Republican-controlled city and state governments are not interested in seeing the program move forward in heavily Democratic neighborhoods. The service programs offered by the EZ initiative may well improve the quality of life for some EZ residents, but it is doubtful that they will be able to compensate for the serious cutbacks in school expenditures, health services, and housing maintenance.

## CONCLUSION

To provide a historic context for these nine case studies, federal policies and their impacts on poor urban neighborhoods are recounted in the next chapter. In addition, urban redevelopment initiatives supported by philanthropic and corporate sponsors are identified.

## NOTE

1. For a contemporary account of the tragedy, see Bellamy (1997).

# CHAPTER 2

# FEDERAL POLICY AND POOR URBAN NEIGHBORHOODS

W. Dennis Keating

## INTRODUCTION

Poor urban neighborhoods in the United States have felt the impact of federal policies for more than six decades. The Great Depression beginning in 1929 and New Deal reforms of the 1930s mark the origin of federal policies that have long affected poor urban neighborhoods. These policies have been both positive and negative in their impact. To understand the past evolution of these neighborhoods and to project their future, it is critical to understand these federal policies. This chapter will trace the evolution of federal policies and review their impact.

Urban neighborhoods grew tremendously in the late 19th century and in the early 1900s in the absence of federal policy. The great migration to the United States and its cities, first from abroad and then from the rural South (largely African Americans), occurred without any federal intervention. Rather, it was economic and social factors—such as the demand for cheap labor in industrializing U.S. cities; racial, ethnic, and religious discrimination and economic problems in the countries and regions of origin; and greater freedom and opportunities—that fueled the growth of U.S. cities in this era. These migratory waves led to the formation of urban villages populated by successive ethnic and nationality groups and, with the advent of the black migration, racial ghettos.

## THE PROGRESSIVE ERA

The residents of these neighborhoods were mostly poor because of their origins. Many were displaced peasant farmers and laborers, attracted by employment opportunities and the prospects of a better life for their children. Their neighborhoods were dense, with the residents unable to afford better housing than cheap, overcrowded tenements. Municipal governments did little to alleviate these conditions until the Progressive movement took hold at the beginning of the 20th century and instituted reforms such as public health regulations, housing codes, and city planning.

Instead of government addressing social problems, it was the settlement house movement that presaged the emergence of social work that attempted to improve the lot of the poor living in these largely immigrant neighborhoods (Fisher, 1994; Halpern, 1995). Typically, the early settlement houses were founded by middle- and upper-class volunteers, often supported by philanthropists. They sought to provide social services and usually shunned advocacy of political and social reforms that might provoke conflict in an era of union organizing and radical activism. The interests of slum dwellers were left to the ward-based and boss-led political machines that then dominated U.S. cities.

## THE LIBERAL NEW DEAL

With the coming of the Great Depression and the New Deal, the relationship between poor urban neighborhoods and the federal government changed, both for the better and for the worse. Massive unemployment and the collapse of the housing and banking sectors, along with the construction industry, hit poor and working-class urban neighborhoods especially hard. Without jobs or savings, homeowners faced foreclosure and tenants faced eviction. The ability of cities to cope with these unprecedented conditions was quickly overwhelmed. After their pleas for emergency federal relief measures were rebuffed by President Herbert Hoover, the mayors of large U.S. cities turned in 1933 to his successor, Franklin D. Roosevelt. Although they did not win federal action on the scale that they requested, the Roosevelt New Deal did initiate programs designed to counteract the depression through government-triggered investment ("pump priming") that would reduce unemployment. As a last resort, the federal government itself funded public works projects employing the unemployed, with the impact felt the most in the nation's cities and poor neighborhoods (Gelfand, 1975).

To stabilize the housing sector, the New Deal reformed the financial institutions critical to home financing and home building. The Federal Housing Administration (FHA), created in 1934, was to become a key factor in the post–World War II suburban housing boom that emptied out many older urban neighborhoods. The FHA played a major role because its underwriting policy, taken in part from the policy of the National Association of Real Estate Boards (NAREB), favored the insurance of new rather than old housing and redlined both those neighborhoods with a racial mix and minority neighborhoods (Jackson, 1985). The combination of the impact of the Great Depression on investment and building in poor neighborhoods, the housing emergency caused by World War II, the FHA's policies, and the pull of suburbs over the 20th century, especially since the 1950s, left many older urban neighborhoods stagnant. Mass foreign migration to central cities ended in the 1920s, although it reemerged for certain groups and affected cities such as Los Angeles, New York, Miami, and San Diego over the past few decades. The great black migration north from the rural South was largely over by the 1950s. Thus, those who deserted the urban villages for the metropolitan suburbs were not being replaced in many cities by new immigrants, as had occurred earlier in the century.

Following World War II, U.S. cities lobbied the federal government for an urban redevelopment program to finance the clearance of slum neighborhoods. The hope was that the New Deal era experiment in slum clearance and public housing could be greatly expanded. The New

Deal began slum clearance as an employment policy. Cleared sites were redeveloped for public uses, including low-rent public housing for those living in substandard housing in poor neighborhoods. Despite fierce opposition from the real estate lobby and conservatives, Congress passed a public housing program in 1937 (Gelfand, 1975). This program was a local option policy. Suburbs, smaller cities, and many cities in the South and West chose not to participate. Public housing became a program largely confined to the major cities of the Northeast and Midwest.

In its early stages, public housing was a vast improvement for poor neighborhoods. In the place of former slum neighborhoods, public housing offered well-built apartments with amenities, including open space, for the working poor. Despite cost restrictions that limited design, location, and services, and despite racial segregation policies, long waiting lists demonstrated the need for public housing. The early promise of public housing was soon handicapped. World War II prevented further construction. After the end of the war, the conservative and real estate industry alliance almost killed the federal public housing program. It barely survived in 1949, and the 1952 election of President Eisenhower severely limited the building of new housing in the 1950s. By 1957, housing reformer Catherine Bauer Wurster, one of the creators of public housing, lamented the dreary deadlock over the program (Wurster, 1959).

Increasingly, the image of public housing became that of the isolated, high-rise barracks ghetto of the very poor, epitomized by the Pruitt-Igoe project in St. Louis that was demolished in 1973. Except for the elderly, little new public housing has been built since the 1960s. Instead, beginning in the 1970s, the federal government switched to housing allowances as the major alternative to public housing. This followed a 1973 freeze of federal housing subsidy programs triggered by the rising costs of federally assisted housing and scandals surrounding privately sponsored but publicly subsidized housing programs (Hays, 1995).

Despite the successes of public housing, it has no political constituency to protect it from the budget cuts that have beset the Department of Housing and Urban Development (HUD) since the Reagan administration cutbacks in federal housing programs of the 1980s. Public housing projects in poor neighborhoods have deteriorated because of the very low income of the occupants and inadequate federal funding for maintenance and modernization. In the 1990s, the HOPE VI program produced some success in the improvement of selected large urban projects, yet Clinton's HUD, fearful of elimination by the newly elected Republican Congress in 1995, proceeded to demolish some of the worst of high-rise public housing, with the promise of replacing lost units with low-rise units and housing allowances.

Despite such reforms, most urban neighborhoods resist the building of any public housing because of its negative image. Tenants eligible for rental assistance often find it difficult to locate apartments in many neighborhoods where landlords do not want them. Thus, millions of very-low-income tenants remain in substandard private apartments because of the inadequacy and underfunding of federal housing programs.

## URBAN RENEWAL

In 1949, while public housing barely survived, Congress easily enacted legislation establishing an urban redevelopment program aimed at revitalizing central cities. Later renamed *urban renewal*, it would become one of the most controversial federal programs to affect urban neighborhoods. Progressive reformers envisioned slum clearance as a way to demolish slums and expand the public housing program to replace the substandard housing that would be removed. That hope was destroyed in the 1949 federal housing legislation. Instead, while substandard housing would be removed, there was no effective mandate to replace these units with similar housing, either on-site or off-site. The occupants of this housing were given the hollow assurance that they would be relocated into replacement housing.

While urban renewal removed hundreds of thousands of low-rent housing units in the 1950s and 1960s, most were not replaced by comparable units. Little urban renewal land was redeveloped for public housing. The housing needs of the displaced poor often were ignored. Eventually, urban renewal became known as "Negro removal" (Greer, 1965). The reason for this pattern was that local governments, in alliance with downtown business interests (characterized as urban regimes and growth machines), designed and implemented urban renewal primarily as a program to revitalize the central business district (CBD) to compete with the burgeoning suburbs (Logan & Molotch, 1987). The aim was to replace slum neighborhoods and obsolete land uses with new retail and office buildings and, where feasible, market-rate housing. Parking garages were built to encourage suburban commuters to work in and visit the downtown, and cities successfully lobbied to ensure that the new federal interstate highway would have beltway connections to the CBDs. This required the further destruction of poor neighborhoods, where land was the least expensive and the residents were least able to win political support to oppose the plans of the urban highway engineers.

Without doubt, the urban renewal and highway programs were the most visibly destructive federal policies to be visited on urban neighborhoods. Eventually, resistance to urban renewal grew. Residents gained legal standing to demand adequate relocation housing. As the costs of urban renewal grew and its benefits were questioned, the debate over its utility grew (Anderson, 1964; Bellush & Hausknecht, 1967; Wilson, 1966). Herbert Gans and Chester Hartman were two of the most eloquent and persistent critics of urban renewal. Gans (1962) decried the destruction of Boston's West End Italian urban village. Hartman fought the Yerba Buena project in San Francisco's South-of-Market neighborhood. The plan for a new convention center, retail and office buildings, and market-rate housing was revised as a result of litigation to include at least some replacement housing for its low-income residents (Hartman, 1974, 1984).

Resistance to urban renewal and the urban riots of 1964-1968 led to reforms in the urban renewal and highway programs. Resident participation in planning and one-for-one replacement of demolished low-income housing were mandated. These reforms came, however, as these programs ground to a halt, in part because of the very opposition that led to these reforms. Urban renewal ended as an independent program in 1974. When the Urban Development Action Grant (UDAG) program was created by the Carter administration in 1978 to promote economic development in distressed central cities, it sought to avoid the problems that overcame urban renewal. Although it did not displace the poor, its record of alleviating unemployment in poor urban neighborhoods was controversial. UDAG funding went primarily to downtown, rather than neighborhood, development. This short-lived program was terminated by the Reagan administration in 1986 as part of its policy of urban aid cutbacks.

## THE 1960S: URBAN RIOTS AND
## AMBITIOUS FEDERAL PROGRAMS

The federal anti-poverty program left some legacies, such as the Head Start and Legal Services programs, beyond its demise in the early 1970s under the Nixon administration. It also left a debate that continues more than two decades later. Defenders of the ideas behind federally sponsored anti-poverty programs argued that their impact had been limited by underfunding and the loss of political support from the Johnson White House, Congress, and city halls. Detractors countered that the so-called war on poverty had been based on misguided theories such as community action and that there was no evidence that it had any structural impact in reducing poverty. This same debate has included other federal programs such as food stamps and housing (e.g., Burton, 1992). Even as the war on poverty quickly lost momentum within a few years of its inception, the nation's attention was focused on the poor neighborhoods of many cities as urban riots became a regular summer occurrence.

Federal attention to the problems of poor urban neighborhoods peaked in the 1960s. The federal government under liberal Democrat President Lyndon Johnson launched a war on poverty in 1964 (Unger, 1996). Although not limited to cities, it captured the most attention in those cities where its Community Action Program set the residents of poor neighborhoods in conflict with institutions identified as oppressive or nonresponsive to their needs, ranging from governmental actors such as police and welfare officials to slum landlords and absentee business owners. Bypassing city halls, the Office of Economic Opportunity (OEO) directly funded community organizations with a stated goal of maximum feasible participation of the poor. After a series of well-publicized confrontations between the poor, aided by federally funded community organizers and Legal Services lawyers, mayors successfully sought to defuse protest politics, which they saw as undercutting their power. Restrictions on participation by the poor, funding cutbacks, and bureaucratic roadblocks brought these efforts to an end (Fisher, 1994). It was later argued that empowerment of the poor was never intended to be a federal policy (Moynihan, 1969).

As poor neighborhoods in the central cities burned, beginning with the Watts uprising in Los Angeles in the summer of 1965, the federal response was dual: police and military action to restore law and order, and efforts to address festering social problems that fueled unrest. The former peaked with the use of federal troops to restore order in Detroit in 1967. The latter featured the Model Cities program, launched in 1967 by the U.S. Department of Housing and Urban Development (HUD), itself created only in 1965, as its first major urban initiative.

The Model Cities program emerged from a presidential task force charged with reviewing urban policies and considering the idea of major federal investment in revitalizing a selected number of central cities (Haar, 1975). The result was a recommendation that HUD coordinate the use of its programs and other federal programs affecting urban areas in targeted poor neighborhoods with the aim of promoting both physical revitalization and providing expanded social services. The hope was to avoid the pitfalls of the urban renewal, highway, and community action programs. The Model Cities legislation was enacted in 1966, but congressional politics expanded its reach from a handful of cities, first to 75 and later to 150 cities. Instead of a significant infusion of additional federal funds into the model neighborhoods in these cities, the diversion of federal priorities and funding into the expanding Vietnam War meant that actual funding for revitalizing poor neighborhoods was modest. With President Lyndon Johnson's preoccupation with Vietnam and civil unrest, and with his decision in March, 1968, not to run again for president, the Model Cities initiative lost momentum. HUD was unable to exert much influence over programs administered by older and larger federal departments. Although Model Cities did survive after the election of Richard Nixon, it was not seen as a major priority and in 1974 it too was terminated.

## A SHIFT TO THE RIGHT: NEW FEDERALISM

In its 1968 report, the National Advisory Commission on Civil Disorders headed by Governor Otto Kerner called for a wide range of federal responses to the social problems of poor neighborhoods and urban ghettos. The Kerner Commission warned against the widening gulf between whites and blacks, the affluent and the poor, and cities and suburbs. With the assassination of liberal Democratic presidential contender Robert Kennedy in June, 1968, and the election of conservative

Richard Nixon over liberal Democrat Hubert Humphrey in November, 1968, the national political climate became much more conservative. Although the Nixon administration, facing a Democratic Congress, did continue many of the Kennedy-Johnson domestic programs (e.g., new subsidized housing programs enacted in August, 1968, on the recommendation of the President's Committee on Urban Problems headed by industrialist Edgar Kaiser), it cut back on their funding and eventually terminated many of them (e.g., Model Cities and the anti-poverty program). This shift to the political right would affect federal policies toward poor neighborhoods for the next three decades. Nixon's urban policy adviser, Daniel Patrick Moynihan, formerly a member of the Kennedy and Johnson administrations, was said to have counseled a federal policy of benign neglect toward cities and their poor neighborhoods.

The Nixon administration sought to end many competitive categorical programs, with the aim of combining them into block grant programs that would be administered by the states. Federal control of funding allocation would diminish because these funds would be allocated by formula and then distributed by state or local governments, rather than the federal government. This was termed the New Federalism.

Even as Nixon was forced to resign, the third of these block grant programs was enacted by Congress in August, 1974. The Community Development Block Grant (CDBG) program combined several categorical HUD programs and allocated annual funding by formula to entitlement cities and, for the first time, urban counties. Congress required that the funds be used to meet such national goals as the elimination of slums and blight, and to benefit principally persons of low and moderate income. The entitlement localities were given considerable discretion as to how they spent CDBG funds.

Despite congressional protection for the previous recipients of such categorical programs as urban renewal and Model Cities, namely central cities, the CDBG formula redirected much of the funding to smaller cities, cities in the South and Southwest, and suburban cities. This was seen as rewarding the growing Republican constituencies in these regions and the metropolitan suburbs. Despite the problems that befell public housing and urban renewal programs, their benefits were aimed at the poor and neighborhoods where the poor were concentrated. In contrast, the CDBG legislation required only that its funds principally benefit the poor, as well as moderate-income, residents of the entitlement communities. This made CDBG much less of a redistributional program (Rich, 1993).

As administered by the Ford, Reagan, and Bush administrations, HUD oversight was deliberately restrained (Hays, 1995). The Carter administration did actively oversee CDBG performance and advocated geographical and social targeting to increase the benefits to low-and moderate-income residents of those neighborhoods receiving CDBG funding. In contrast to the Johnson anti-poverty and Model Cities programs, however, the CDBG program was never touted as a federal effort to eliminate slums or seriously address poverty in blighted neighborhoods. Its funding and impact have waxed and waned over more than two decades, but CDBG funding priorities generally have not been targeted locally to the poorest neighborhoods or the neediest residents. Instead, the tendency of local governments has been to spread the benefits as widely as possible (Rich, 1993).

## HUD AND THE NEIGHBORHOODS

Poor neighborhoods had never been directly represented by federal agencies or programs, even though they were often the target of those programs, for better and worse. The Carter administration was the first and only one to change this. It created an office of neighborhoods within HUD and appointed neighborhood activist Rev. Geno Baroni to head it. To assist the neighborhood development organizations or community development corporations that were beginning to grow in the United States, the Carter administration enacted a Neighborhood Self-Help Development (NSHD) program to provide grants to these organizations to promote neighborhood revitalization. Launched in 1980 but with only minor funding, this program was later evaluated as a success (Mayer, 1984).

To address broader urban revitalization issues, the Carter administration appointed a National Commission on Neighborhoods in 1977, another innovative policy initiative. A divided commission, however, produced a massive list of new federal programs that would have cost far more than envisioned by the Carter administration. As a result, its recommendations were ignored. With Carter's defeat in the 1980 presidential election, his modest urban and neighborhood initiatives were quickly reversed.

The Reagan administration sought to reduce the role of the federal government by eliminating many of the social programs that benefited poor neighborhoods or, as an alternative, reduce and redirect their funding. The rule of the private market was to replace an interventionist federal government in determining urban policy (Clarke, 1984). Reagan's HUD eliminated the office of neighborhoods and the NSHD program. HUD's budget was the hardest hit of the federal cabinet departments as Reagan's Office of Management and Budget reduced federal social spending. Democratic control of the House, however, prevented the total elimination of such programs as CDBG and subsidized housing, including public housing. The Reagan administration resisted calls for federal action to address urban problems, including homelessness. A Democratic Congress did enact a modestly funded demonstration neighborhood assistance program and in the 1990 housing act reserved some subsidized housing funding for community-based nonprofits in recognition of their efforts to build and rehabilitate low- and moderate-income housing in poor neighborhoods (Vidal, 1992).

## LENDERS AND INVESTMENT IN POOR NEIGHBORHOODS: HMDA AND CRA

The FHA decision, at its inception in the 1930s, to follow the practices of Realtors and lenders in redlining poor and minority urban neighborhoods in its housing insurance underwriting policies had as devastating an impact on these neighborhoods as did the federal urban renewal clearance program on those neighborhoods later demolished for redevelopment and for highways. With the decline of these neighborhoods following World War II, as the income of remaining residents declined, as better-off residents left for the suburbs (often enjoying the benefits of FHA-insured newly built housing and federally subsidized highways) and as the manufacturing base of central cities declined, housing grew older and fell into disrepair, and public services and safety declined. Lender redlining of these neighborhoods meant either unavailability or prohibitive rates for home purchase and home repair loans, home insurance, and personal and business loans. Lender redlining was both a result of these trends and a contributing factor.

As urban decline accelerated beginning in the 1960s, many institutional lenders also literally left these neighborhoods, closing their branch banks. This starkly symbolized the plight of these neighborhoods, starved for credit for basic economic activities. In the midst of the mid-1960s urban riots, the newly created HUD forced the FHA to relax its insurance underwriting standards in order to insure mortgages in high-risk inner-city neighborhoods. In many cities, especially with the creation of new lower-income home ownership programs in 1968, the FHA became a major influence. Unfortunately, in all too many cities, FHA-insured financing paved the way for a new wave of white flight and a rapidly accelerating rate of mortgage default and foreclosure by inexperienced home buyers unprepared and marginally qualified financially for home ownership. Fraud by unscrupulous Realtors and, in some cities, by FHA appraisers contributed to the perceived failure of this approach, although this was not the case, for example, in Milwaukee, where the FHA provided counseling for qualified buyers in these neighborhoods (Hays, 1995, pp. 113-121; Squires, 1992).

Chicago epitomized these patterns. It has long been one of the most racially segregated cities in the United States; it is where racial covenants were invented. It was University of Chicago sociologists like Ernest Burgess and Homer Hoyt who justified the policy of redlining based on denigration of black and ethnic neighborhoods. It was in Chicago that the struggle for reinvestment in these neighborhoods began, starting in the West Side neighborhood of Austin in the late 1960s. This led to the formation of the West Side Coalition to fight bank redlining. Its campaign culminated in a pioneering agreement between the Bank of Chicago and community organizations in which this large commercial bank agreed to give priority in its lending practices to targeted Chicago neighborhoods (Pogge, 1992). This, in turn, led to a national organization based in Chicago called National Peoples Action (NPA), led by activist Gale Cincotta, that campaigned for national policy reforms.

What the anti-redlining forces sought was federal policy to require federally chartered and regulated lending institutions to lend in poor and blighted inner-city neighborhoods that previously had been redlined. This met with strong opposition from the powerful lending lobby of commercial banks and savings and loan associations (S&Ls). They opposed any federal mandates that might affect their lending policies, arguing that their fiduciary obligation was to their depositors and investors to earn the highest possible return allowed and that such a mandatory investment policy in poor inner-city neighborhoods might jeopardize their return on investment. Moreover, lenders continued to dispute the allegation that they had engaged in redlining and racially discriminatory lending practices. They claimed that other factors explained the low rates of mortgage lending in inner-city neighborhoods, such as high risk, low rates of application, low appraisals of housing values, and the poor credit records of many applicants, especially lower-income minorities. In Chicago, data on the geographic location of mortgage loans made by savings and loans regulated by the Federal Home Loan Bank Board in the period 1971-1973 made available to the Metropolitan Area Housing Alliance contradicted these lender claims (Squires, 1992, p. 10).

The NPA then pushed for national legislation based on this model. In 1975, Congress enacted the Home Mortgage Disclosure Act (HMDA), which requires federally regulated banks, savings and loans, and credit unions to report annually the number and dollar value of mortgage loans that they make by census tract within the metropolitan areas that they serve. In the 1989 savings and loan bailout legislation, the HMDA requirements were extended to mortgage bankers, who play a significant role in mortgage financing in inner-city neighborhoods. In addition, this legislation requires all reporting institutions to disclose the race, gender, and income of all mortgage loan applicants, along with the final disposition of the applications (Squires, 1992). As Squires (1992, p. 11) notes, a major limitation of HMDA, particularly prior to these amendments, was the absence of any information on demand or the decision-making process that might explain reasons for investment patterns. The HMDA nevertheless did provide community groups fighting redlining with important, if incomplete, information about lending patterns that they were challenging.

The NPA also spearheaded the lobbying that resulted in passage of the Community Reinvestment Act (CRA) in 1977. NPA and its allies sought again to impose lending mandates on lenders, especially S&Ls,

proportional to savings from neighborhood residents. The opposition of the lender lobby defeated this idea again. Instead, the CRA states that federally regulated financial institutions have a continuing and affirmative obligation to help meet the credit needs of local communities in which they are chartered, consistent with safe and sound operation of these institutions (Squires, 1992, p. 11). To comply with this broad requirement, lenders must identify their service areas, identify the credit needs of these areas, and explain how they are meeting their obligation. Federal lending regulatory agencies (e.g., the Federal Reserve Bank) are required to assess the CRA compliance of lenders and take this into account if they apply for permission to merge and expand, which often involves the proposed closings and consolidation of neighborhood branches.

Because federal regulatory agencies have not aggressively enforced CRA, it has been left to community groups to do this, using the opportunities provided by proposed mergers and expansions to file CRA complaints. As a result, lenders faced with such criticism and eager to avoid costly delays and litigation, in cities like Atlanta, Boston, Chicago, and Detroit, have agreed to make large, multiyear commitments to expand their credit in poor, inner-city neighborhoods (Squires, 1992). Squires (1992, p. 12) estimates that $18 billion in CRA-generated reinvestment commitments were made in 70 cities through the early 1990s. These CRA settlements were spurred by Federal Reserve Board studies in Atlanta and Boston that documented continuing patterns of racial discrimination in mortgage lending, a much-disputed finding (Ambrose, Hughes, & Simmons, 1995).

This infusion of reinvestment, combined with various federal subsidy and tax credit programs (e.g., the low-income housing tax credit) and investments made by private corporations and philanthropic foundations (Keyes, Schwartz, Vidal, & Bratt, 1996), has been critical to the efforts of community development corporations (CDCs). It is the CDCs that have been vital actors in the long-term efforts to revitalize inner-city neighborhoods since the urban riots of the 1960s (Keating, Krumholz, & Star, 1996; Vidal, 1992). Following the Republican takeover of Congress in 1994, conservatives threatened to eliminate or seriously weaken the provisions of HMDA and CRA. If successful, this would have undermined much of the efforts of CDCs unless lenders suddenly agreed to reinvestment voluntarily. Led by the National Community Reinvestment Coalition, these proposals were defeated in 1995 and 1996. In addition, the Federal National Mortgage Association (FNMA) announced a major initiative to expand its secondary mortgage markets in several central cities, and HUD unveiled a program to promote lower- and moderate-income home ownership zones in inner-city neighborhoods.

## THE 1992 LOS ANGELES RIOT AND EMPOWERMENT ZONES

The April 29, 1992, acquittal of the Los Angeles police officers indicted in the beating of Rodney King sparked the worst unrest in Los Angeles since the Watts riot in August, 1965. Several days of mob violence resulted in 58 deaths, 2,300 injuries, the destruction of 1,150 structures, and damage to approximately 10,000 small businesses, mostly in South Central Los Angeles (Gale, 1996). Unlike the Detroit riot of 1967 or the riots following the assassination of Dr. Martin Luther King, Jr., in April, 1968, the 1992 Los Angeles riot did not immediately produce a significant federal response (Baldassare, 1994). Seeking reelection and dependent on conservative political forces insisting on the balancing of the federal budget and cutbacks in domestic social programs, Republican President Bush deplored the devastation and did nothing. He refused to sign emergency aid and enterprise zone measures enacted by the Democratically controlled Congress shortly before the election and then vetoed them after his defeat (Gale, 1996, pp. 112-133). It was left to his successor, Bill Clinton, to produce a belated federal response.

Los Angeles did organize Rebuild LA, intended to promote mostly private corporate reinvestment in South Central Los Angeles to assist small businesses and to assist in job training and development for the unemployed. Rebuild LA was unable to raise much of the pledged funding amid a major economic recession in Southern California, and its impact was minimal (Baldassare, 1994; Gale, 1996, p. 11).

Elected as a New Democrat dedicated to moving the Democratic Party more to the political center, Bill Clinton did not make urban policy a priority. While funding for HUD was increased under the leadership of Secretary Henry Cisneros, former mayor of San Antonio, the Clinton administration lost its major domestic reform—health care. The Clinton administration nevertheless pursued an urban aid program similar to the enterprise zone proposal that had been proposed by the Reagan administration and had been blocked by congressional Democrats until the 1992 legislation vetoed by Bush. The Clinton administration proposed the designation of 100 enterprise communities (ECs) (65 urban) and 10 empowerment zones (EZs) (6 urban). These areas were to contain poverty, unemployment, and general distress, similar to those areas designated in the 1960s for the anti-poverty and Model Cities programs and in the 1980s for neighborhood UDAG assistance. When selected competitively, these EC/EZs would be eligible for direct federal grants for social services ($100 million for EZs and $3 million for ECs, over a 10-year period). In addition, employers within these areas would be eligible for federal tax credits for the training and employment of residents. This proposal was included in the controversial 1993 budget bill, which barely passed Congress (Gale, 1996, pp. 134-138).

After Democrats lost control of Congress in the Fall 1994 election, Clinton became increasingly conservative in his policies. His administration, however, pursued the empowerment zone program, announcing the winners of the competition in December, 1994. The six urban empowerment zones were Atlanta, Baltimore, Chicago, Detroit, New York, and Philadelphia-Camden. Despite this being viewed as the delayed response to the aftermath of the 1992 South Central riot, Los Angeles was not one of the six. It was, however, designated as a Supplemental Empowerment Zone, along with Cleveland. This entitled Los Angeles to $125 million in assistance and Cleveland to $90 million, both over 10 years (Gale, 1996, pp. 139-141).

Overall, Congress authorized only $4 billion over a 10-year period (1995-2005) for the EC/EZ program. Compared to the enormous problems confronting the winners in this competition, this is a pittance. In the context of the Clinton administration's agreement to balance the federal budget by 2002, as demanded by congressional Republicans (who contributed to Clinton's reelection as a centrist in 1996), there is little prospect for significantly increased federal aid to cities and distressed central city neighborhoods. Rather, many congressional Republicans called for the elimination of HUD. HUD Secretary Cisneros was forced to reinvent HUD in 1995 to save it from this fate, only three decades after its creation. Cisneros proposed to continue to downsize HUD's staff and to decentralize many of its programs that were not already covered by the CDBG program (Gale, 1996, p. 203). Public housing is to be privatized, with Section 8 certificates replacing many units being demolished. HUD's role in the EC/EZ program is largely one of monitoring the efforts of the local governments involved, similar to the CDBG program. The first assessment concluded that in its early stages, localities were pursuing economic development and employment strategies as proposed, with considerable citizen participation (Nelson A. Rockefeller Institute of Government, 1997). Overall, it was too early to assess the actual impact of this initiative in most of the assisted communities. This is true in most of the case studies that follow.

## CONCLUSION

What then can be said of federal policy as it has affected poor urban neighborhoods over the past six decades? Its impact has been varied, both beneficial and destructive. Its magnitude and direction have been affected by national and municipal politics and by external events. Certainly, federal aid has peaked during perceived periods of economic or social crisis. The two most important examples are the Roosevelt New Deal programs of the Great Depression of the 1930s and the Johnson Great Society programs during the era of urban riots (1964-1968). Presidents Franklin Delano Roosevelt and Lyndon B. Johnson epitomized liberal Democratic politics after their landslide electoral victories in 1936 and 1964, respectively. In contrast, during periods of relative prosperity, social tranquility, and conservative Republican administrations, there has been little constituency for federal aid to cities or to distressed urban neighborhoods. This was exemplified in the administration of Dwight Eisenhower (1953-1961).

The reelection of President Richard Nixon in 1972, his 1973 moratorium on HUD programs, and their subsequent reevaluation signaled a reversal of the 1960s era of federal urban initiatives, however relatively small their magnitude and flawed their implementation. Nixon, followed by Ford, Carter, Reagan, Bush, and Clinton, proposed to reduce the federal role by emphasizing decentralization and block grants rather than categorical federal aid programs. After 1973, Nixon sought to curtail increases in HUD spending. President Ronald Reagan carried on Nixon's policy, as did George Bush, and actively sought to minimize HUD's role in favor of privatization of many of its programs and draconian cuts in its spending authority. Democrats Jimmy Carter and Bill Clinton, both former Southern governors, proved to be fiscal conservatives unwilling or unable to greatly expand HUD's role and funding. Clinton's accession in 1996 to Republican demands for a balanced budget and welfare reform is likely to result in greatly reduced federal aid to the impoverished residents of the poorest urban neighborhoods in the United States. There is no national constituency for mounting large-scale and costly efforts to deal systemically with inner-city problems. Rather, cities have had to try to find substitute sources for declining federal urban aid programs. Federal entitlement programs are being reduced in benefits or eliminated. Federal responsibility for many social programs (e.g., welfare) is being delegated to the states. Poor inner-city residents and their representative organizations are being forced to search elsewhere than Washington for needed assistance.

## URBAN REDEVELOPMENT INITIATIVES

Philanthropic foundations and enlightened individuals have taken the lead in trying to promote innovative ways of dealing with the overwhelming problems confronting distressed urban neighborhoods. The most prominent examples are the community development corporations (CDCs) and community building initiatives.

CDCs trace their origins to the war on poverty, when the late Senator Robert Kennedy intervened to assist fledgling CDCs, beginning with the Bedford-Stuyvesant Restoration Corporation that he helped to form in Brooklyn. CDCs grew in the following three decades, despite the drastic cutbacks in funding for community organizations and HUD housing programs, upon which many relied, during the Reagan and Bush administrations (Keating, Krumholz, & Star, 1996). As Vidal (1997) notes, the community development movement represented by CDCs has matured as an industry. In addition to support from the federal and local governments, CDCs have been provided with critical financial assistance by "intermediaries" such as the Ford Foundation's Local Initiatives Support Corporation (LISC) and the Enterprise Foundation created by the late James Rouse to promote the development of community-based low-income housing with corporate support (Keyes et al., 1996). CDCs of varying capacity and with differing histories play a role in the attempted revitalization of most distressed urban neighborhoods, including those highlighted in this book.

These same philanthropic intermediaries have also led an effort to build community. By this, they meant the comprehensive, long-term attempt to revitalize severely distressed urban neighborhoods, combining a variety of resources and promoting resident leadership. Two examples are noteworthy, among many under way.

In 1990, Rouse's Enterprise Foundation joined with the City of Baltimore and its first black mayor, black churches, and community groups to try to transform the historic African American neighborhood of Sandtown-Winchester. Located near downtown, it had experienced a long decline. In 1990, its population of 10,000 mostly poor African Americans was only one-third of its peak population. A Community Building in Partnership (CBP) was formed to address the neighborhood's myriad problems over a decade. Planning groups developed action plans for community building, education, human services, and physical and economic development, with a comprehensive plan produced in 1993. By 1997, CBP could report many successes (Goetz, 1996; Walsh, 1997), although achievement of its long term-goal remained a formidable undertaking. The Sandtown-Winchester process served as a model for Baltimore's Empowerment Zone program.

In 1991, former President Jimmy Carter conceived of The Atlanta Project (TAP), an ambitious 5-year initiative intended to bring together corporate leaders, volunteers, and community residents to develop an anti-poverty urban agenda for the city. The key strategy was to organize around 20 neighborhood "clusters" in metropolitan Atlanta. In 1996, TAP ended formally, although it left behind some still-existing organizations (Smith, 1997).

The evaluations of TAP's achievements are mixed (Walsh, 1997). Certainly, the many problems facing Atlanta's poor neighborhoods remain, as exemplified by the example of Peoplestown in this book. Community building initiatives nevertheless stand out as an alternate approach to past reliance on federal urban aid programs, which themselves are now much reduced.

# CHAPTER 3

# ATLANTA

*Peoplestown—Resilience and*
*Tenacity Versus Institutional Hostility*

Larry Keating

## EARLY HISTORY

Peoplestown is a small (2,527 persons), poor (officially 50% poverty), African American (95%), predominantly rental (85%), well-organized, politically tenacious, indigenously redeveloping neighborhood located one and one-half miles south of Georgia's capital. Developed as a streetcar suburb after the Atlanta Electric Railway built a line along Capitol Avenue in 1885, the neighborhood initially consisted of middle- and upper-class Victorian homes, shacks, and one- and two-room homes. Typical of Atlanta and the South, the neighborhood contained both racially integrated sections and an enclave of exclusively black residences. Along the streets and avenues with rail or streetcar service (Ormond, Capital, and later Atlanta), and west of Capital Avenue on Washington Street and Crew Streets, Victorian homes for upper- and middle-class whites were constructed. These streets were both wider

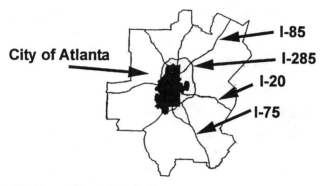

**Figure 3.1.** The Metropolitan Atlanta Region
SOURCE: Map prepared by Caitlin Waddick.

than the streets in the interior of the neighborhood east of Capital and
paved. Alleys divided each block and provided separate access for the
black households who lived in the small shacks at the rear of many lots.

South of Atlanta Avenue and east of Capital Avenue, narrower unpaved streets and smaller lots mark the general location of a black
enclave. The Peeples family, possible namesakes of the neighborhood,
owned 66 of the lots in this section of the neighborhood as it was
beginning to develop in 1894. By 1925, insurance maps showed 550
residences, of which 510 were single-family homes and 40 were in
duplex structures. Because homes in the black section of the neighborhood were not shown, these figures represent only the white and
internally integrated portions of Peoplestown.

The 1920s and 1930s also brought significant change to the white
portions of Peoplestown. Victorian era residential development had
compactly encircled the city's central business district along one-to
two-mile-long radial street railway and trolley lines, but when the
upper classes began to commute in automobiles, they also began to
move north along Peachtree Street. By the late 1920s, upper-class sub-
divisions had been built as far as four and one-half miles north of the
center of the town. Middle-class development followed, particularly on
the east side of the Peachtree axis, where there were fewer industrial
uses. As a consequence, demand for the less prestigious southside
streetcar suburbs declined.

The 1939 Works Progress Administration (WPA) survey provides the
first clear picture of the black enclave on the east side of the neighbor-

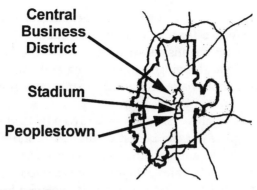

**Figure 3.2.** The City of Atlanta
SOURCE: Map prepared by Caitlin Waddick.

hood—12 blocks were almost 100% black. The proportion of renters in this section was 33%. Overall, the neighborhood had 1,159 housing units, with 792 (68%) rental units, 367 (32%) owner-occupied units, 188 (16%) black-occupied units, and 971 (84%) white-occupied units.

The next reliable data point regarding the racial composition of the neighborhood is 1960 census block data. By then, the number of housing units had increased to 1,509, of which 751 (54%) were black-occupied, 639 (46%) were white-occupied, and the rest were vacant. The neighborhood was still a predominantly rental neighborhood, with 961 (63.7%) rental units and 429 (28.4%) owner-occupied units. Whites still lived on the west side of the neighborhood and blacks on the east side. The total population was 6,831.

## EXPRESSWAYS AND NEIGHBORHOOD DESTRUCTION I

The late 1950s marked an apex in Peoplestown's development. During the next 30 years, expressway construction, urban renewal, Model Cities, and stadium construction undermined the tenuous stability the lower-middle-class and poor neighborhood had attained.

Planning began in 1940 for the expressway system. The Central Area Improvement Association (CAIA), precursor to the present-day Central Atlanta Progress (CAP) as the primary institution representing central area businesses, successfully lobbied the city and the Georgia Highway Department for the retention of consultants to prepare plans for expressways linking the downtown with the expanding suburbs in 1944. The initial plan called for the primary north-south expressway to circumnavigate the central business district (CBD) on the west side, where it would traverse a declining manufacturing and warehouse district. This alignment would have brought the expressway south out of the downtown area west of Peoplestown.

The location of highways, however, is driven by land use as well as transportation goals. At the behest of the CAIA and the Hartsfield administration, the location of the downtown connector was shifted to the east side of the central area, where it could also accomplish the tasks of removing parts of three low-income black neighborhoods, where it could separate the central business district area from two public housing communities, and where it could serve as a buffer between the central area and the residual portions of the black neighborhoods.

The realignment moved the route of the expressway south of the central area to a path occupied by Peoplestown. In the early 1960s, Interstate 75/85 replaced the western edge of the neighborhood, demolishing 110 primarily single-family homes, separating Peoplestown from the Pittsburgh neighborhood, and setting the stage for a sequence of local urban renewal, Model Cities, and stadium construction programs.

## URBAN RENEWAL

Although the initial damage to Peoplestown from the expressway was an indirect consequence of local transportation and land use policies directed at other poor black communities, the urban renewal program intentionally brought the same set of local policies to Peoplestown's borders. The Washington-Rawson urban renewal project implemented the city policy of demolishing low-income black neighborhoods surrounding the CBD in the southeast quadrant of the intersection of the east-west and north-south expressways, three blocks north of Peoplestown.

The initial 1957 plan and the legal justification for the Washington-Rawson project was to replace more than 1,000 low-income black-occupied housing units with moderate-income ownership housing and light industry, businesses, schools, and parks. Half of the 598 acres to be cleared were to be housing, and the other half were to provide jobs, recreation, and education for area residents. Part of the rationale for the project was to redevelop the area as a buffer between the increasingly low-income black neighborhoods of Peoplestown and Summerhill and the CBD. In addition, an informal (and illegal) agreement between the Atlanta Housing Authority (then the city's redevelopment agency) and local real estate interests not to construct public housing on cleared renewal land sought to implement a third goal of local land use and redevelopment policy: minimizing public housing near the CBD.

In the spring of 1963, plans for the area changed abruptly. According to then Mayor Ivan Allen, Jr., he spontaneously conceived of a stadium on the urban renewal land to convince Kansas City Athletics owner Charles O. Finley not to end his secret visit to the city with diminished impressions of Atlanta as the location for his major league baseball team. A more prosaic and probably accurate version of the story says Allen located the stadium at the Washington-Rawson site to thwart black community proposals for black public housing there. Stadium planning and design began within 6 weeks of Finley's visit as Citizens and Southern Bank president Mills B. Lane provided nearly one-half million dollars in front-end money, while having himself been named treasurer of the Atlanta-Fulton County Recreation Authority, the legal entity designated to build and manage the stadium. Coca-Cola Bottling Company executive Arthur Montgomery was named chairman of the Authority. In subsequent planning, the privately controlled Recreation Authority off-loaded parking for 35,440 fans to the Summerhill, Peoplestown, and Mechanicsville areas. The 1964 feasibility study for the stadium proposed only 4,100 parking spaces. The Milwaukee Braves, who agreed in February, 1964, to move to Atlanta, negotiated the provision of an additional 2,500 parking spaces by the city within 10 years of their 1966 arrival.

The stadium had a capacity of 55,000 people. When the stadium was full, 12,462 cars had to find parking. After 1976, the 6,600 formal, public sector spaces left the other 5,862 cars to find parking in the surrounding residential areas. The land use consequence was that the first two to three blocks surrounding the stadium had almost all their houses burned down or demolished for informal sector parking. The band of vacant land encircling the stadium proved to be immune to development and a barrier to development in the interior of the three neighborhoods for more than 25 years.

The event transportation plan equally disregarded Summerhill, Mechanicsville, and Peoplestown's interests. The plan stipulated that all major streets in the three neighborhoods be converted to one-way streets carrying traffic into the stadium area for one and one-half hours prior to a game and that they be converted to one-way streets carrying traffic away from the stadium for more than an hour after a game. To leave their neighborhoods before a game going against the flow of traffic and the wrong way on a one-way street, local residents describe a process of trying to guess which police would not pursue them. Reaching their homes from outside the neighborhood after a game simply was not possible in a car.

The third negative impact on the adjacent neighborhoods has been late-night noise and fireworks. Efforts to institute curfews, or limitations on how late games can last or when fireworks can be ignited, have been opposed by the baseball team and rejected by the city government.

Urban renewal demolition extended as far south as the north side of Georgia Avenue and eliminated the hospital/clinic, the grocery store, the ice cream parlor, and the butcher that had served Peoplestown, Mechanicsville, Summerhill, and parts of the Pittsburgh, South Atlanta, and Capitol Homes neighborhoods for more than 25 years. Finally, the destruction of more than 1,000 low-income housing units coupled with a citywide replacement housing policy that demolished at least 14,000 more low-income units than it replaced exacerbated substandard housing conditions by pushing displaced residents into Peoplestown and surrounding neighborhoods.

The combination of destructive local land use policies, suburbanization of the Atlanta Jewish population, school integration, and racial fears had ended Peoplestown's 80-year history as an integrated neighborhood by 1970. That year's census showed 99% of the homes to be black-occupied.

The most publicized riot in Atlanta in the 1960s occurred on the Tuesday after Labor Day in 1966, on Peoplestown's northern border and centered at Capital Avenue and Ormond Street. Housing conditions headed the list of grievances neighborhood representatives reported to the mayor in the riot's aftermath.

In the spring after the riot, Emmaus House, an agency of the Episcopal Diocese of Atlanta, was founded as a settlement house at Capital Avenue and Haygood by the Reverend Austin Ford. Emmaus House operates a community action center that has helped catalyze organizing in Peoplestown and political action on Model Cities, highways, stadium construction, local tax policy, and the death penalty.

On the surface, following the riot and belated recognition of damages done by urban renewal, local policy appeared to shift toward rehabilitation, neighborhood representation, and conservation of existing areas. The Allen administration set a goal of 17,000 new units of low- and moderate-income housing at a Mayor's Conference on Housing in November, 1966. Relocation housing and repair of existing housing were identified as priorities. Peoplestown, Summerhill, and Mechanicsville were included in the new Model Cities program area, and grassroots participation was promised.

Promises, however, exceeded capacities and political will. The Model Cities plan targeted the district south of Georgia Avenue and north of Peoplestown as one of three Economic Growth Cores intended to produce high quality commercial and residential development. Over consistent neighborhood opposition emanating from Emmaus House and the Poverty Rights Office, which consistently picketed the Model Cities office and in November, 1970, conducted a "sleep-in" in the Model Cities director's office, the area was cleared. It first became the locus of most of the 2,500 stadium parking spaces promised to the baseball team, then in 1991, the proposed site of the Olympic Stadium. The Model Cities program destroyed the remaining commercial facilities on the south side of Georgia Avenue.

## PEOPLESTOWN ORGANIZES

In contrast to the handpicked neighborhood representatives characteristic of urban renewal and Model Cities program, Emmaus House's community organizing strategy focused on self-determination and authentic and widespread participation. In addition to more easily organized homeowners, Emmaus House organized renters and youth. Columbus Ward, who would later lead the neighborhood through the Olympic Stadium fight and who is president of the Peoplestown Revitalization Corporation (PRC), was president of the Senior Teenager Group in 1970 at the age of 15.

Although low- and moderate-income housing production never approached the levels initially proposed, Peoplestown residents did gain some new housing. Eighty-seven units were built using the 221(d)(3) and 221(d)(4) programs, just north of the neighborhood on Washington and Crew Streets. Within the neighborhood, the Emmaus House Housing Authority acted as the nonprofit partner (with a 0.1% ownership interest) for two Section 236 (later converted to Section 8) rental apartments.

## EXPRESSWAYS AND NEIGHBORHOOD DESTRUCTION II

While Emmaus House and Peoplestown worked to strengthen the neighborhood internally in the wake of renewal, Model Cities, and highway demolition during the 1970s and 1980s, external and autonomous decisions continued to exert substantial influences on Peoplestown's prospects. Protracted and ultimately successful opposition to a proposed inner loop expressway through gentrifying neighborhoods on the city's near east side resulted in substitution of the reconstruction and widening of I-75/I-85 on the west side of Peoplestown. The Georgia Department of Transportation (GDOT) widened the expressway and added two or three auxiliary entry, exit, and interchange lanes along Peoplestown's border. The reconstruction also effectively eliminated the direct stadium access/egress interchange at Georgia Avenue, thereby requiring much of the stadium traffic to travel through either Peoplestown or Summerhill. Demolition of approximately 50 residences and 15 businesses in Peoplestown was required. Finally, the project eliminated several hundred event parking spaces between Peoplestown and the stadium. None of the substantial land use impacts on Peoplestown was analyzed by the City Bureau of Planning.

Grassroots resistance to the widening focused on the homes and businesses scheduled to be demolished. Emmaus House, Ethel Mae Mathews, and some of the residents of the McDaniel-Glenn public housing development in Mechanicsville pressed for minimal relocation and equitable treatment of relocatees. This resistance did achieve a small reduction in the total number of units to be demolished.

The expressway widening completed in the mid-1980s cost Peoplestown four blocks as the western neighborhood boundary was shifted two blocks east. In addition, the two western block faces along Pulliam between Ridge and Atlanta were effectively lost to the neighborhood because expressway noise, pollution, and insufficient buffering rendered 24 of the 27 lots vacant by 1996. Thirteen of the 27 lots are presently significantly tax delinquent.

Reorganization of the interstate interchanges away from direct access to the stadium forced more event traffic through the neighborhood, and the loss of formal parking to the expressway increased pressure for informal parking lots to push deeper into the northern section of Peoplestown. Some of the parking eventually was replaced by the GDOT, but because it was replaced in less accessible areas across the expressway, pressure for informal sector parking in Peoplestown was not relieved. The GDOT proposed replacing the parking on the east side of the expressway closer to the stadium with decked parking, but this solution was successfully resisted by the Atlanta-Fulton County Recreation Authority.

## THE OLYMPIC STADIUM

On August 1, 1986, consultants submitted conceptual site and stadium designs to the governor for new football and baseball stadiums. The Atlanta Falcons' threatened move stimulated the study. The Falcons eventually received a new football-only stadium on the west side of the CBD, but among the alternatives proposed by design firm Hellmuth, Obata & Kassabaum was a new baseball stadium, which was both the first public acknowledgment that the Braves were also seeking a new stadium to replace the 20-year-old Atlanta-Fulton County Stadium and the first public exploration of where that stadium would be located. Together with affiliated parking, the new facilities would have demolished 15 blocks of Summerhill east of the stadium. Ominously for Summerhill, all four of the alternative plans for the Atlanta-Fulton County Stadium area envisioned placing a new stadium east of the present one in Summerhill, and each alternative required significant demolition in the neighborhood (Hellmuth, Obata & Kassabaum, 1986).

Clarence Stone extensively documented the current governance of Atlanta by a coalition of the largely white business corporate elite and black elected officials. The mutually beneficial relationships the two groups provide one another both enable them to mobilize the significant political and economic resources required to undertake large projects and inhibit the thwarting of their combined wills. Operating out of public view, the business elite/black elected official regime decided during 1987 and 1988 to pursue the hosting of the 1996 Olympics in Atlanta.

## SUMMERHILL'S DEAL WITH THE REGIME

The internal dynamics of Atlanta's Olympic bid preparation process is mostly secret, but during the 4 years ending in September, 1990, when the city was selected to host the Olympics, the following events occurred.

1. A new baseball stadium, primarily financed by Olympic revenues, was incorporated into the bid.
2. Central Atlanta Progress (CAP) and other businesses began fund-raising for the Washington-based Urban Land Institute (ULI) to prepare a redevelopment plan for Summerhill.
3. Summerhill, with city concurrence, extended its official neighborhood boundaries east to incorporate the stadium and the parking lot to the south.
4. Douglas Dean, the primary spokesperson for Summerhill, was selected by the Atlanta Organizing Committee (AOC) to be one of two "community representatives" on the AOC.
5. The location of the new stadium was shifted to the south of the existing stadium, away from Summerhill and into Peoplestown's borders.
6. Summerhill and Dean backed the new stadium in a publicly acknowledged exchange for immunity from stadium-affiliated new parking in Summerhill and business and government backing for their redevelopment plan.

Public confirmation of an agreement between the regime and Summerhill to shift the stadium toward Peoplestown in exchange for Summerhill endorsing the stadium has not been made, but the circumstantial evidence strongly suggests that it was integral to the agreement.

Douglas Dean said, "We have a dream that a community, destroyed by a stadium 25 years ago, will once again thrive with the beginning of a new stadium—the Olympic Stadium. . . . By 1996 we ought to have a show piece in Summerhill" (Walker, 1990, p. E1). Columbus Ward, who was the elected chair of the Neighborhood Planning Unit (NPU) that included Summerhill, Peoplestown, and Mechanicsville, said "The deal Summerhill made to support the stadium put everyone in jeopardy" (Hiskey, 1991, p. D2).

During the 6-year buildup to the Olympics, Summerhill became the showcase of Atlanta's business elite, city government, and Olympic organizers' low-income neighborhood redevelopment efforts. City and private funds for the ULI plan were raised, and the plan was developed. Summerhill's expansion of its boundaries to include the old and new stadia withstood challenges, the agreements on no new parking in Summerhill were honored, Dean continued to praise the stadium, $66.4 million in public and private funds were invested in the neighborhood, 189 new ownership houses were constructed, a 64-unit public housing community was converted to cooperative ownership, and 130 owner-ship units began development. Four of Atlanta's banks, two of the most prestigious law firms, Coca-Cola affiliated foundations, hometown corporations Delta Airlines and Home Depot, and the city's largest multiple donor foundation each contributed significant amounts to Summerhill's redevelopment. Summerhill captured 72.6% of the public and private money invested in the six neighborhoods surrounding Olympic venues (Keating, Creighton, & Abercrombie, 1996).

Summerhill's deal politically marginalized Peoplestown and Mechanicsville's opposition to the stadium, threatened to shift all the new public parking into the two neighborhoods, threatened to drive informal sector parking deeper into each community, and made them subject to more of the noise and fireworks impacts of a new stadium. Their opposition to both the private process that selected the site and the unmitigated potential impacts cut them off from the fiscal and political resources the city, the business elite and the Olympic orga-nizers assigned to Summerhill.

## PEOPLESTOWN MOBILIZES OPPOSITION

When the Olympics and the stadium were announced, Columbus Ward, Ethel Mae Mathews, Duane Stuart, and other neighborhood residents immediately formed Atlanta Neighborhoods United for Fairness (A'NUFF) to resist the stadium proposal. The group called for a publicly accessible planning process to determine if a new stadium was needed. (The fact that Georgia Tech's 45,000 seat Bobby Dodd Stadium was not scheduled for any Olympic sporting events called the need for a new stadium into question.) If a new stadium was needed, A'NUFF called for environmental impact analyses of multiple potential sites. It also called for sufficient parking in decks, for buffers to protect adjacent communities, for a publicly accessible planning process to organize traffic and parking, and for recognition and respect of the fact that each neighborhood had redevelopment plans in place.

A month after A'NUFF's formation, CAP sent its only black executive to meet with stadium area residents. Vice President John Leak told the residents that the stadium would be built regardless of their opposition. Referring to the power his corporate backers had in civic affairs, Leak said, "We have to realize that when a certain movement occurs in the community with the support of a certain level of leadership, some things are going to happen, some things are going to change" (Roughton, 1990, p. A1).

Two weeks later, the editor of the *Atlanta Constitution*'s editorial pages interpreted A'NUFF's opposition to a second stadium as much more politically threatening opposition to the Olympic Games (Teepen, 1990, p. G7). Ethel Mae Mathews and Duane Stuart retorted that they were not opposed to the Olympics but only to the stadium. They responded,

> This trivializes the intent of those who are against the building of the second stadium. The issues here are not so simplistic. Twenty-five years of urban renewal, stadium building, and highway construction have served to maintain this area as an underdeveloped enclave in the shadow of downtown. This is not a matter of a selfish, cranky few who oppose "progress," but a matter of many in a community resisting the ongoing destruction of a poor working class and predominantly African American community. (Mathews & Stuart, 1990, p. A12)

During the more than 2 years required to design and finalize the details for the stadium, A'NUFF demonstrated at Olympic organizer Billy Payne's suburban home, convinced the Atlanta Planning Advisory Board (composed of the chairs of the 24 NPUs) to endorse and support their platform, organized an alliance with the Atlanta Labor Council, participated in the short-lived (6 weeks) and ultimately aborted Atlanta Committee for the Olympic Games (ACOG)-organized stadium area

Neighborhood Task Force, and precisely researched the potential impacts of the stadium.

As the irrevocability of the Olympic stadium decision became clearer, A'NUFF backpedaled to press for sufficient parking within the confines of the existing stadium and parking areas (which would have required decked parking), enforceable bans on informal sector parking within the neighborhoods, an event transportation plan, community improvement funds deriving from parking revenues, and limitations on late-night games and fireworks.

In 1991, the Community Design Center of Atlanta (CDCA), a nonprofit planning and architectural technical assistance provider with whom I work, conducted the analysis of the event parking situation around the stadium referred to earlier. When informed that the deficit in public spaces was 5,862 and that 12,462 total spaces were needed, Columbus Ward argued that the total should be 13,500 to provide some security for the neighborhoods in case travel behavior changed, a clear example both of the necessity of planning advisers to work closely with the communities they are advising and of the more complete perceptions of neighborhood conditions residents bring to planning processes.

Ward began lobbying in neighborhood councils for 13,500 spaces. The neighborhoods could agree that sufficient parking be provided only as long as the new spaces were in decks or underground. When in July, 1992, the Recreation Authority and the Braves decided on surface parking—even the insufficient number of 10,000 spaces—the decision split the neighborhoods because approximately 2,500 new surface spaces would have to be located in the neighborhoods. At least one of the three neighborhoods (not Summerhill and less likely Peoplestown) was going to lose. A unified front arguing for 13,500 spaces was no longer politically possible.

## THE CONSEQUENCES OF OPPOSITION

Also in July, 1992, the mayor and the commissioner of planning and development presented the "Olympic Development Program" prepared by city staff to the International Olympic Committee (IOC) in Barcelona. Detailed budgets were presented to the IOC for Summerhill, Vine City, Techwood, and Mechanicsville, but Peoplestown had only a "costs to be determined" entry. The message was not lost on Columbus

Ward and the Peoplestown leadership. "The City periodically tried to pressure us into dropping our resistance to what they were doing," Ward said of the omission (personal communication, Columbus Ward, December 14, 1993).

Voters in Atlanta approved a $149 million bond issue for infrastructure repair in July, 1993, and some of the proceeds were allotted to the Corporation for Olympic Development in Atlanta (CODA) for CBD and neighborhood improvements. CODA prepared a redevelopment plan for Summerhill in 1993 and oversaw the completion of plans for two other neighborhoods during the following 2 years. CODA and the city made the preparation of a plan under CODA's auspices a prerequisite for most categories of city funding and access to some of the bond proceeds. Peoplestown had prepared a plan in 1991, but it was not prepared by CODA; therefore, Peoplestown was ineligible for most forms of city funding. Because Peoplestown's CODA plan was not begun until April, 1996, the pot was empty, and the Olympics had come and gone by the time Peoplestown could meet the new criteria.

Peoplestown's opposition to the preemptive deal that the adjacent Summerhill neighborhood cut with the Atlanta regime to construct the new Olympic stadium (now the Atlanta Braves' Ted Turner Field) on Peoplestown's northern border cut the neighborhood off from government and financial resources targeted at more compliant poor communities in the buildup to the 1996 Olympics. Activists from A'NUFF opposed the location of the new stadium, the offloading of substantial unmitigated externalities (parking, event traffic, highway and infrastructure reconstruction, and noise) onto either unbuffered contiguous poor neighborhoods or the public treasury, and the secrecy and inaccessibility of the decision-making process. Political opposition did not succeed in blocking the stadium or in reversing private deals about public facilities, but Peoplestown did succeed in protecting some of its own interests: The stadium site was moved one-quarter mile north, away from the neighborhood border; a community improvement fund deriving from stadium parking revenues generates $300,000 annually, of which Peoplestown receives one-third; no new public parking was located within the neighborhood; informal sector parking was ostensibly banned by local ordinance, which the PRC continues to press the city to enforce; and finally, the PRC purchased the key parcels of land along the border with the stadium and is developing them in a way that protects neighborhood interests and meets resident aspirations.

Although cut off from most corporate philanthropy and public resources before the Olympics, the PRC still managed to acquire and rehab
one Resolution Trust Company (RTC; a federal agency created to dispose of properties acquired by federal insurance agencies as a consequence of the savings and loan crisis of the 1980s) rental property (20
units), to purchase and demolish a crack-infested irredeemable rental
complex (104 units), and to force the rehabilitation of two Section 8
properties in which it indirectly holds a minuscule interest. PRC's
integrity, competence, and persistence have attracted substantial funding from the Enterprise Foundation (a national foundation that invests
in and leverages investments in nonprofit housing) and political respect
from others in the post-Olympic period. Ironically, Peoplestown appears to have traversed the frenetic Olympic preparation period in a
better position than its complicit neighbor. Whereas Summerhill's gentrification now threatens its predominantly poor population, Peoplestown continues to try to directly improve conditions for the majority
of the neighborhood who are poor, and it is beginning to make progress.

During the March, 1993, final negotiations, Peoplestown managed to
beat back a city proposal for a 351-space lot on Washington Street in the
neighborhood. Earlier opposition to the location of two cooling towers
on the neighborhood's borders succeeded in moving them away from
the neighborhood.

## INDIGENOUS REDEVELOPMENT

During the 1980s, working out of Emmaus House and the Poverty
Rights Office, a small cadre of Peoplestown residents fought against the
most damaging aspects of highway expansions and stadium parking
intrusions into residential areas. In addition, Emmaus House and the
residents organized summer youth programs, medical referrals, emergency financial assistance, social services client advocacy, after school
youth programs, and multiple programs for seniors. In 1990, prior to
the Olympics/stadium announcement, Columbus Ward, Gene Fergeson, Ethel Mae Matthews, and several other Peoplestown residents
added community-based action to foster redevelopment to the existing
social service programs. One of the first steps engaged the CDCA to
work with the community to prepare a redevelopment plan. Initially
focused on housing and strengthening community organizations, the

plan shifted emphasis when the Olympic stadium was announced. The threat of the new stadium and its land use impacts absorbed much of the organizers' energy over the next 7 years, but they also galvanized the community to craft and initiate development projects in the northern section of the neighborhood to counter the threatened construction of event parking lots and supplant informal sector parking.

Peoplestown Revitalization Corporation, the community development corporation residents formed in 1991, acquired strategically located properties along Ormond to secure the northern border of the neighborhood against intrusive uses. King Manor, a partially occupied, dilapidated, crack-infested 104-unit apartment complex, was demolished and the land acquired with assistance from the Enterprise Foundation, the Metropolitan Atlanta Olympic Games Authority, HOME funds, and the Atlanta-Fulton County Land Bank Authority. Ninety Low Income Housing Tax Credit apartments are under construction. The PRC has gained control of most of the property across the street on Capitol Avenue and is in the late planning stages for a development of 34 townhouse units and a small, neighborhood-oriented retail complex. Along Washington Street, the PRC rehabilitated and manages 20 rental units acquired from the Resolution Trust Corporation.

In the early 1990s, Peoplestown leadership used the Emmaus House Housing Authority 0.1% equity position in Boynton Village and Capitol Vanaria as leverage to force the general partners to rehabilitate and fully occupy both Section 8 complexes. Along Washington Street, the PRC worked with MAOGA and Habitat for Humanity to build 20 Habitat homes before the Olympics.

After the city failed to adopt and enforce effective prohibitions on informal sector overflow parking from stadium events, the PRC organized indigenous enforcement patrols that have used blockades and other innovative approaches to preventing the intrusions. The focus, however, has not been exclusively on the north end of the neighborhood.

Using the Land Bank Authority's capacity to acquire title to tax delinquent property, the PRC now owns and is redeveloping 11 units in the old black enclave section of the neighborhood. Farther south in the Grant Terrace section, the PRC has packaged nine 203-k loans with Empowerment Zone second mortgages to forestall the deterioration of a small subdivision. The PRC worked with the City of Atlanta's Housing Commissioner to channel the city's expanded, pre-Olympic demolition program to 40 units throughout the neighborhood that were beyond redemption. The PRC also channeled Christmas in April, a local volunteer group, to the cosmetic rehabilitation of 20 owner-occupied units.

## FUTURE PROSPECTS

Future prospects include both threats and opportunities. On the positive side, the $100,000 annual revenue from Braves-controlled parking, while not fully compensatory for the damage the stadium triggered, constitutes base funding for the PRC and some Emmaus House social service programs. PRC administrative support from Atlanta Neighborhood Development Partnerships has been relatively steady and should remain so in the near term. Capacities to complete projects solidified after the PRC negotiated a steep learning curve in developing its first few projects. The completion of the CODA redevelopment plan just after the Olympics finally opened a gate to potential expanded city funding.

Many of Atlanta's CDCs have opted, like Summerhill, for a gentrification strategy that, when successful, will increase the housing problems of poor people. The PRC board of directors instead has been steadfast in its focus on building and rehabilitating rental housing that is accessible to the substantial majority of neighborhood residents who are poor. Finally, the struggles of the buildup to the Olympics have strengthened and broadened the base of the PRC as an organization.

Prospective threats are the conversion of project-based Section 8 certificates at Boynton Village, Capitol Vanaria, Capitol Avenue, and Capitol Towers to tenant-based certificates, which could lead to a reduction in the number of low-income units in and near the neighborhood. If the erosion of federal housing support continues, residents of these buildings will be at risk. The stadium parking issue remains unresolved. Peoplestown's vigilance has blocked the most damaging proposals and will have to be maintained. Event traffic also remains a substantial problem. A transportation management plan has been promised and could, with effective community participation, ameliorate some of the most severe effects.

Successive groups of low-income black residents have fought tena-
ciously to build a stable community in Peoplestown. The first settlers
constructed the nucleus of their community on the unpaved roads of
the enclave and in the shacks lining the alleys behind white residences.
Grace Barksdale and other residents overcame Atlanta's institutional
racism in the 1920s and 1930s to obtain electricity, gas, and street paving.
Henry Phipps participated in those struggles, and he and his wife led
the fight for a school in the 1940s and 1950s. From the 1960s through the
present, Ethel Mae Matthews, Columbus Ward, Gene Fergeson, Duane
Stuart, and others have fought nearly continuous battles against the
destructive effects of expressways, urban renewal, a stadium, Model
Cities, expressway widenings, and a second stadium. The character and
composition of the forces that suppressed Peoplestown's development
have changed from the deeply embedded racism of the Victorian era,
to the only slightly less virulent racism of the 1920s and 1930s, to
consciously destructive white business elite strategies using expressways, urban renewal, and Model Cities to eliminate poor black neighborhoods in the 1950s through the 1980s. Most recently, the coalition of
black elected officials and the largely white business elite coopted the
adjacent poor black Summerhill neighborhood to provide political
cover to build the Olympic stadium on Peoplestown's border. Had
Peoplestown not aggressively fought this compact, the stadium would
be on top of the neighborhood, and much of the remainder of the
community would be a parking lot.

One hundred years of struggle have left the neighborhood impoverished and damaged by the noxious land uses that surround it, but still
resiliently fighting. The post-Olympic threat is to continue to resist the
intrusive by-products of the stadium—formal event parking is still
insufficient, informal sector parking is tenuously controlled by neighborhood vigilance, and a long-promised event traffic management plan
has yet to be formulated or to produce enforceable mechanisms to
protect Peoplestown. Post-Olympic opportunities are to build on the
reinvigorated commitments of residents, the strategically focused plans
and development activities of the PRC, and the respect produced by the
last 9 years of principled political action.

To be continued . . .

# CHAPTER 4

# CAMDEN, NEW JERSEY

*Urban Decay and the Absence*
*of Public-Private Partnerships*

Robert A. Catlin

## INTRODUCTION

In his 1989 book *Unequal Partnerships: The Political Economy of Urban Redevelopment in Postwar America*, Gregory Squires questions whether or not the public-private partnerships developed in central U.S. cities after World War II have really worked to the advantage of most citizens. Nationwide observations during the 1980s and early 1990s are mixed. In Boston, Sacramento, and Portland, Oregon (Dreier, 1989; Smith, Guagnano, & Posehn, 1989), the results seem to be positive for all concerned. In Pittsburgh (Sbragia, 1989), Detroit (Thomas, 1989, 1995), Milwaukee (Norman, 1989), and Louisville (Cummings, Koebel, & Whitt, 1989), some benefits trickled down to low- and moderate-income citizens, but the clear winners were large corporations and upper-income individuals. In Houston, (Feagin, Gilderbloom, & Rodriguez, 1989), the only beneficiaries were big business; low- and moderate-

51

income residents saw an absolute decline in the amount and quality of affordable housing and public services.

Camden, New Jersey, is unique: It stands out as a city where no one profited. Despite its proximity to Center City Philadelphia and a benevolent state government, a complete lack of leadership and public-private partnerships has meant that *nobody* profited: not corporations, institutions, or citizens at any level. Camden, once a prosperous industrial city, is now a ward of the State of New Jersey.

This chapter examines how Camden, a city of considerable promise up until the end of World War II, arrived at such a sorry state. Camden's fate is similar to many other mid-sized "rustbelt" cities such as East St. Louis, Illinois; Gary, Indiana; Bridgeport, Connecticut; and Flint, Michigan (Catlin, 1993; Nelson & Meranto, 1977). Lacking *Fortune* 500 corporate headquarters, major businesses, and major universities, Camden has become a victim of squabbling politicians, uninterested enterprises, and an indifferent and frustrated state government bureaucracy that had turned from benevolently liberal in the late 1980s and early 1990s to rigidly conservative by 1997.

## BACKGROUND

Camden began in the late 17th century as a farming community providing staples such as pinewood, melons, and pork sausage to Philadelphia via ferryboat. After the Civil War, the farming and commercial city enjoyed its first form as an emerging industrial city. Railroads linked Camden to Trenton, Newark, Jersey City, and the rest of the nation. Ferry lines linked Camden to Philadelphia across the Delaware River. Newly developing industries included steel, woolen goods, lumber products, and chemicals. In 1869, the Campbell's Soup Company began operations in Camden. RCA Victor and New York Shipbuilding began their tremendous growth in 1890. By 1880, Camden was the 44th largest city in the United States. As the number of jobs and people grew, the future looked bright.

Despite the economic boom, however, the city was plagued by crime and racial tensions. Camden was castigated in the press as a city "worse than Philadelphia in its wickedness" (Kirp, Dwyer, & Rosenthal, 1995, p. 18), reflecting on political corruption, labor strife, and crime. Race

**Table 4.1**  Camden, New Jersey, Population Change by Race and Hispanic
Origin[1], 1900-1994

| Year | Total Population | % Change | % White | % Black | % Hispanic Origin and Other[2] |
|---|---|---|---|---|---|
| 1900 | 75,935 | — | 98.5 | 1.4 | — |
| 1910 | 94,538 | +24.5 | 98.4 | 1.6 | — |
| 1920 | 116,309 | +23.0 | 92.5 | 7.3 | — |
| 1930 | 118,700 | + 2.1 | 90.3 | 9.6 | — |
| 1940 | 117,536 | –1.0 | 89.2 | 10.6 | — |
| 1950 | 124,555 | +6.0 | 78.6 | 21.4 | — |
| 1960 | 117,159 | –5.9 | 64.3 | 30.6 | 5.1 |
| 1970 | 102,551 | –12.5 | 53.4 | 39.1 | 7.5 |
| 1980 | 84,910 | –17.2 | 27.7 | 53.0 | 21.3[3] |
| 1990 | 87,492 | +3.0 | 13.3 | 56.4 | 30.3[3] |
| 1994 (est.) | 90,000 | +2.9 | 10.5 | 55.0 | 36.8[3] |

SOURCE: U.S. Census.
1. Until 1960, persons of Hispanic origin were counted as either black or white depending on the enumerator's subjective judgment.
2. Hispanic origin or other less than 1% until 1960. "Other" category was 1.0% in 1980, 1.5% in 1990, and 1.8% in 1994 estimate.
3. Hispanic origin could be either black or white.

relations were a continuing problem even before the "great migrations" beginning in 1915. Although blacks had lived in Camden before the American Revolution, they were the object of discrimination. From the Civil War until the end of the 1960s, African Americans in Camden were systematically segregated and discriminated against in public accommodations, housing, education, and employment by public policy and private practices.

An important event in Camden's economic growth was the opening of the Benjamin Franklin Bridge in 1926. The bridge connected Camden with Center City Philadelphia and provided for pedestrians, cars, trucks, and railroad cars. It helped expand the city's commerce with new hotels, department stores, rail terminals, and oceangoing vessels from an expanded port.

World War II also increased the city's industry as Campbell's Soup, RCA Victor, and the shipyards led by New York Shipbuilding reached record employment levels. The city's population increased from 117,536 in 1940 to 124,555 in 1950. After World War II, industries continued to expand. First Lady Mamie Eisenhower christened the first nuclear submarine, the USS *Nautilus*, at New York Shipbuilding in 1953. RCA introduced the United States' first color television set in 1955 and also enjoyed its first billion dollar year in gross sales. Campbell's Soup recorded record sales during the 1950s, and during that same period, the shipyards employed up to 35,000 workers, sometimes building eight ships at once.

Economic problems for Camden began to surface at the same time as prosperity peaked. A series of protracted strikes in the late 1940s prompted some companies to close or move to the suburbs. The ferries, unable to compete with the Benjamin Franklin Bridge, terminated service in 1952. Three years later, the daily *Courier Post* built its new plant in suburban Cherry Hill, where burgeoning shopping malls were weakening downtown Camden. The major convention hall went up in flames in 1958. More important, by 1960 the Federal Housing Administration (FHA) had redlined all of Camden as unacceptably risky, and mortgage money dried up almost completely. By 1960, Camden's population had decreased from its 1950 high of 124,555 to 117,159, and 1970 saw a further decrease to 102,551. Meanwhile, population in suburban Camden County increased by more than 100,000 between 1950 and 1970. Camden's economy weakened between 1950 and 1970. Half of the city's manufacturing jobs were lost, 22,000 in all, while manufacturing jobs in the region increased from 80,000 to 197,000 during the same time period. In the early 1960s, two main Camden industries, the Esterbrook Pen Company and the New York Shipbuilding Company, shut down, putting thousands out of work.

During the 1960s, plans were formulated for new rail mass transit lines and highways to connect Camden's suburbs with Philadelphia, literally bypassing the city altogether. In 1969, the Philadelphia-Camden subway, which had terminated in downtown Camden, was extended to the suburbs of Collingswood, Haddonfield, Cherry Hill, and Lindenwold. Known as the Hi-Speed Line, this transit facility was state of the art and became a prototype for heavy rail systems later built in San Francisco, Washington, D.C., Atlanta, and Baltimore (Delaware River Port Authority, 1996). Also planned during the 1960s was Interstate Highway 676, which linked suburban Camden County directly to the Benjamin Franklin Bridge. Because of legal difficulties, it was not completed for several years.

The response of Camden's leadership was a massive urban redevelopment plan geared to bring back suburbanites and jobs. Initiated in

1965, the plan called for 280 acres of waterfront land to be cleared for a shopping center, high-rise residential complexes for workers in nearby downtown office buildings, and entertainment facilities. This "city within a city" would then be linked to the suburbs by the Hi-Speed rail line, bypassing Camden's blighted inner-city neighborhoods, which were becoming more black and Latino (see Table 4.1). Proposed by Jerry Wolman, a real estate developer who owned the Philadelphia Eagles professional football team, the plan ignored the realities of economic decline, suburban white flight, and the prevailing disinvestment on the part of banks and other lending institutions (Derthick, 1972). The plan assumed that it was possible to cordon off a sizable part of the city and pretend it was immune from the decay that ate away at its borders. This "city within a city" never did generate support from Camden's private sector elites, and the developer went bankrupt before he could assemble financing for redevelopment.

By the late 1960s, Camden's African American community, led by the NAACP, CORE, and a local group known as the Black People's Unity Movement (BPUM), assailed the redevelopment plans as "Negro removal." In 1970, these groups, aided by the Camden County Regional Legal Services, filed suit against all urban renewal and highway construction projects. The suit known as *Coalition vs. Nardi*, the city's mayor, was upheld in court because of Camden's total lack of an adequate relocation plan. By 1973, Mayor Joseph Nardi agreed to modify the plans to provide several hundred units of new housing for the poor. In turn, the Camden County Regional Legal Services dropped their lawsuit. Meanwhile, racial tensions erupted in a major riot in 1971 that burned 20 city blocks to the ground. Twelve people were killed, 23 were injured, and more than 200 families were made homeless.

Although the agreement between Mayor Nardi and the activists resulted in the construction of a 23 story high-rise building for low-income citizens known as Northgate and the training of several black and Puerto Rican workers from Camden as building tradesmen, it was a Pyrrhic victory at best. In January, 1973, President Nixon suspended all federal housing and redevelopment assistance programs. This action virtually killed all new assisted housing projects. Stung by the 1971 riots and the *Coalition vs. Nardi* compromise, Camden's private sector retreated from urban redevelopment in the city and confined its activities to suburban ventures. In 1973, Mayor Nardi was replaced by Angelo

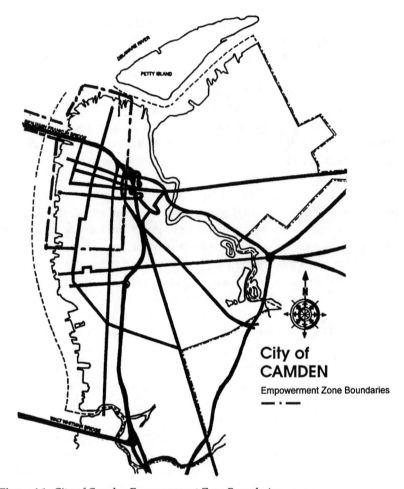

**Figure 4.1.** City of Camden Empowerment Zone Boundaries

Errichetti, who, realizing the increasing political power of the city's African American community, brought several black leaders into city government. Still, Camden's problems intensified.

In 1975, more violent crime was recorded in Camden than in any other city of its size in the nation. By 1980, Camden's population declined to 84,910 from the 1970 level of 102,551, and in that same year, Interstate 676, the highway that bulldozed almost 2,000 housing units occupied mainly by poor blacks and Puerto Ricans, was completed. In

1981, Camden elected its first African American mayor, Randy Primas. Like Richard Hatcher in Gary, Indiana (1967) and Kenneth Gibson in Newark, New Jersey (1970), he inherited the proverbial "hole in the doughnut."

## MAYOR RANDY PRIMAS TRIES TO REBUILD CAMDEN

By 1980, Camden was 53% black, 21% Hispanic origin (mainly Puerto Rican), and 28% white. By 1983, not only did this city have a black mayor, but it had a Latino city council as well. For the first time in Camden's history, the governing body, mayor, and city council reflected the minority racial makeup of its residents. Although this spoke well for minority group political aspirations, Mayor Primas and his follow-ers were beset with problems remaining from the benign neglect of the past. They were worsened by the gradual withdrawal of assistance by the city's business community. By 1980, Camden's population had fallen to its lowest level since 1900. Only 10,000 manufacturing jobs remained, compared to 45,000 in 1960 and 22,000 in 1970. More than 6,000 units of housing had been either bulldozed by scattered renewal projects, demolished by Interstate 676, or abandoned as the white middle- and working-class population fled Camden for the suburbs. In 1980, Camden's unemployment was twice the national average, and almost 50% of its people were living below the poverty line. Primas's election could not have taken place at a worse time, as President Ronald Reagan reduced the budget of the Department of Housing and Urban Development (HUD) by 80% during the 1980s. At the same time, a national shift from an industrial to a service economy left cities such as Camden literally in ruins (Caraley, 1992; Nenno, 1989; Peterson & Lewis, 1986).

In 1981, Thomas Kean was elected governor of New Jersey by a margin of only 2,000 votes out of more than two million ballots cast. Kean, a liberal-moderate Republican, in his two terms spearheaded a movement to revitalize New Jersey's economy by putting billions of state dollars into transportation, infrastructure, and education projects. Realizing that distressed cities such as Newark, Jersey City, and Cam-den needed state assistance to offset major losses in tax ratables, the Kean administration offered to subsidize Camden's government, but at a price. In exchange for state aid that by 1990 would amount to one quarter of the city's budget, Camden was forced to accept projects that no other municipality in south New Jersey would take. The first was a 600-bed maximum security state prison located on the Delaware River just north of the Benjamin Franklin Bridge. The prison's highly visible location would dampen development interest in the city. The second project was an upgrade to the city's sewage treatment plant. Completed in 1987, the enlarged complex began processing 55 million gallons of waste generated every day by all of Camden County's 500,000 residents. It replaced 46 local treatment plants shut down when suburban residents articulated concerns about the degraded environmental quality in their own communities. The third undesirable project forced on Camden was a solid waste facility that burns trash from all the Camden suburbs. Completed in 1989 and located in the center of Camden's residential area just off Interstate 676, this facility is serviced every day by more than 200 trucks driving through Camden, bound for the incinerator. When the wind blows toward the surrounding housing, the stench is overwhelming.

The Kean administration did bring some positives to Camden. In 1985, the state helped to create the Cooper's Ferry Development Corporation, Camden's first formal public-private partnership. This non-profit entity was given a mandate to develop the remaining Camden waterfront with a variety of public and private uses. With support from the Cooper's Ferry Development Corporation, a $55 million state aquarium was built on the Camden waterfront south of the Benjamin Franklin Bridge and the state prison. This facility opened in 1992 and was followed by a new marina, a parking garage, and an outdoor entertainment facility known as Sony-Blockbuster. This complex has proven to be a qualified success, although its economic impact on Camden has been slight.

During his term of office, Mayor Primas followed the directives of the Camden County Democratic Party machine. This was in contrast to Richard Hatcher of Gary, Carl Stokes of Cleveland, Kenneth Gibson of Newark, and Coleman Young of Detroit, all of whom were the first African Americans to be elected mayor of their cities but who followed a path clearly independent of their party machines once elected to office (see Greer, 1979; Rich, 1989). When Primas left office in 1989, the Camden County machine backed Aaron Thompson as his replacement.

Thompson was seen as a safe and compliant person. After Thompson took office, however, he began to criticize the Cooper's Ferry Development Corporation and private businesses in Camden for failing to share jobs and contracts with minority firms. In response, Campbell's Soup, which had closed its Camden factory in 1989, throwing 1,000 employees out of work, gradually reduced their Camden-based corporate staff, transferring many white-collar front office personnel to New York City. Campbell's also backed out of an arrangement to relocate corporate headquarters to the Camden waterfront. When Thompson came up for reelection in 1993, the machine refused to back him and remained neutral. A bitter Democratic Party primary in June 1993, saw a loose confederation of African Americans, Puerto Ricans, and some whites rally around Camden School Board superintendent Dr. Arnold Webster. Webster defeated Thompson for the Democratic Party nomination and, with Democrats outnumbering Republicans 20-1, won easily in the general election.

## THE ADMINISTRATION OF DR. ARNOLD WEBSTER

Arnold Webster was a Ph.D. who became Camden's superintendent of schools in 1987. After winning election as mayor, Dr. Webster retired from the Camden school system, earning a pension of $65,000 annually. As he was leaving the school system, Dr. Webster wrote himself a check for $83,000 in unused vacation pay. Although perfectly legal, the action was not good form for the leader of this impoverished city.

Webster inherited a city with daunting problems but considerable opportunities. The 1990 census showed that Camden had actually gained about 2,500 residents during the 1980s, with most of the newcomers being Latinos from Puerto Rico and other Caribbean islands. That population was generally poor, with almost 60% living below the poverty level; almost 70% percent of those in poverty were children or elderly. Fewer than 100 households reported incomes of $50,000 or more. The average residential property value was $25,000, compared with the statewide average of $200,000, and Camden's entire assessed value was worth less than one prime piece of Atlantic City real estate (Kirp et al., 1995, p. 181). On the other hand, the city seemed on the brink of possible revival with the new waterfront development under way. At the same time Dr. Webster was elected mayor, Christine Todd Whitman was elected governor of New Jersey. A moderate Republican who favored abortion rights and affirmative action, Whitman promised a 30% state tax cut in her campaign. When it was implemented in 1994-1996, the tax cut created revenue shortfalls that caused local property taxes to rise. Whitman continued to provide state aid to Camden while keeping a watchful eye on Camden's fiscal practices.

Dr. Webster soon found out that running a city was much more complicated than leading a public school system. Mayor Webster was not included in 1994 negotiations involving a new basketball arena on the Camden waterfront. When he complained bitterly to the press that he was being ignored—and, after all, the proposed arena was in *his* city—the *Courier Post* replied, "someone tell Arnold not to worry; when the governor is ready she will tell him all about it" (Parillo, 1994). Eventually, in 1996, the arena was built in Philadelphia, but the point was made: When it came to major projects in Camden, others were to call the shots, not the city of Camden's bankrupt government.

Mayor Webster then severed all ties with the Cooper's Ferry Development Corporation (Rouse, 1996). Because his administration was not aligned with the Camden County Democratic Party organization, Webster essentially painted himself into a corner. For the next 3 years, he found his government virtually completely outside the development decision-making loop. Off-year elections between 1993 and 1996 produced a city council hostile to Mayor Webster. Although Webster, with cooperation from Mayor Edward Rendell of Philadelphia, was able to obtain $120 million in November 1994, for a bistate Empowerment Zone including parts of Philadelphia and Camden, he found it difficult to move ahead. Two years later, Camden was still without a plan to spend its $21 million share of Empowerment Zone money, and the federal government threatened to cut off Camden's funding and shift the monies to the other nine cities with Empowerment Zones[1] (Mendics, 1996).

A series of scandals beset Webster's administration between 1994 and 1997. First, the city's parking authority was taken over by the state because of repeated deficits caused by mismanagement. In the first year of state supervision, the authority posted a surplus (Moore, 1995). The city's housing authority went into receivership, and HUD threatened to take back $10 million in urban development grants previously pledged to Camden (Angeles, 1994). In 1995, Camden led the nation in per capita murders, and the resultant negative publicity forced Robert Pugh, the city's first African American police chief, to "retire" in August 1996, but only after Pugh, a resident of suburban Washington township, paid himself more than $100,000 for vacation time, overtime, and sick leave (Riordan, 1996b). In February 1996, the State of New Jersey's Department of Community Affairs blasted Camden's fiscal practices by releasing a 300-page audit that accused city government of mishandling its finances. City government was also attacked by a group of nonprofit housing sponsors, the Camden Churches Organization for People (CCOP), for failing to develop a list of almost 4,000 vacant houses, many of which were not even boarded up (Riordan, 1996a).

Mayor Webster's response to the mounting wave of public criticism was a July 11, 1995, unveiling of "Project Arizona." Project Arizona was a loose collection of small, older, urban redevelopment projects, most of which were located within the Empowerment Zone. The most promising were restoration of the *Moshulu*, a former restaurant ship then rusting and idle on the Delaware River, and the opening of a branch of the famous Harlem-based soul food restaurant Sylvia's. One year later, no progress had been made on any of the Project Arizona developments. The *Moshulu* was transported across the Delaware River to Philadelphia, where, by 1997, it had been rehabilitated and was set for reopening. Sylvia's was unable to obtain needed permits and gave up plans for the restaurant (Riordan, 1996c). As of the summer of 1997, urban redevelopment in Camden was at a standstill both within and outside the Empowerment Zone.

## HOW CAMDEN MIGHT IMPROVE
## ITS RECORD IN URBAN REVITALIZATION

Camden is a city with tremendous revitalization potential. Located just across the Delaware River from Center City Philadelphia and connected to that city by bridges, a rail mass transit line, and interstate highways, as well as having large parcels of abandoned industrial sites with all utilities in place, Camden would be an ideal location for "back office" functions of major Philadelphia corporations, storage and warehousing facilities, and a variety of uses on the waterfront facing the Philadelphia skyline. According to the New Jersey State Department of Environmental Protection, less than 20% of the total abandoned industrial acreage has "brownfield" problems. Gradual abandonment of the city by business has hurt revitalization efforts. Unlike some successful revitalized cities, Camden does not have a major research university or large medical center within its borders or immediate metropolitan area. On the other hand, nearby Pennsylvania cities including Allentown, Bethlehem, Wilkes-Barre, and Scranton also lack "placebound" major universities, multiple *Fortune* 500 corporate headquarters, or large medical centers, but they have been able to achieve revitalization success. What Camden might do to boost not only physical urban revitalization but also the overall quality of life for its 90,000 citizens is to utilize the following approaches:

1. Promote strong, effective local governmental leadership;
2. Develop a comprehensive planning approach to set the stage for revitalization;
3. Move toward at least a form of metropolitan government; and
4. Encourage direct transfers from federal and state government and the private sector to nonprofit organizations.

Since the 1970s, older postindustrial central cities with majority minority populations have, for the most part, elected articulate, forceful leaders who, while not necessarily enjoying support among the business and institutional community, were at least able to articulate a vision for their city. By doing so, they rallied the vast majority of citizens to their side. This produced a force that had to be taken seriously by federal and state government and even the local private sector. Included in this group of mayors were Richard Hatcher of Gary, Coleman Young of Detroit, Maynard Jackson and Andrew Young of Atlanta, Dr. Lionel Arrington of Birmingham, and Kenneth Gibson and Sharpe James of Newark.

Camden's local governmental leadership since the 1970s has not been able to rally even the mass of citizens to its side. Errichetti was caught up in an elaborate Federal Bureau of Investigation sting operation known as ABSCAM and had to step down as mayor in 1981 (Kirp et al., 1995, p. 182). Primas, under withering criticism for permitting the state to place a prison on the waterfront, went from being Camden's mayor to heading a state agency under Democratic Governor Jim Florio (Kirp et al., 1995, p. 186). Thompson, a pleasant, well-meaning individual, worked best in private situations, and Dr. Arnold Webster turned off business leaders, state and federal government, and even Camden citizens with his well-publicized pension and vacation pay arrangements. Camden governmental administrators also lacked credibility with masses of citizens. The police chief resided in the suburbs. The superintendent of schools (also a suburbanite) was regularly criticized by state government, the press, and a majority of school board members for low pupil test scores, a higher than average percentage of expenditures for administration, and seven relatives on the school board payroll ("What's Wrong With Camden's Schools," 1996).

Finding articulate, well-educated leadership for Camden will be difficult. Since 1990, the black middle class, especially younger upwardly mobile recent college graduates, has been fleeing Camden for the suburbs. This sense of frustration is best explained by Jonathan Kozol (1991) in his conversation with a black teacher who had just moved his family out of Camden after his house was burglarized. The teacher explained, "I am not angry. What did I expect? Rats packed tight in a cage destroy each other. I got out. I do not plan to be destroyed" (p. 142).

Camden cannot even begin to develop a successful revitalization strategy or means to implement it without a chief elected official painting the vision. Solid leadership is a necessary first step.[2]

Under mayors Primas, Thompson, and Webster, Camden failed to use comprehensive planning as the foundation for a successful revitalization effort. The city's last comprehensive plan was a 1977 document prepared by the well-respected, Philadelphia-based firm of Wallace, McHarg, Todd associates. Hopelessly out of date, that document cannot possibly be used as a backdrop for redevelopment decisions. Although Camden has a capable planning staff, the political agenda has focused exclusively on project development masquerading as planning. There is a "plan" for the waterfront promoted by the Cooper's Ferry Development Corporation. The City of Camden Redevelopment Agency has no less than 23 separate projects and a plan for each one, but it completely lacks an overall redevelopment scheme that ties all these projects together and sets priorities. This focus on projects has produced endless conflict as proponents for various developments fight for their piece of the action, with no one, including city government, coordinating the process with a comprehensive plan to fit all the pieces together. Until such a plan is produced and adopted, revitalization efforts will be compromised.

David Rusk (1993) and other scholars advocate metropolitan consolidation as a means of cost-effective governance. Presently, Camden County's 500,000 residents are balkanized into 56 separate municipal corporations, the smallest of which contains only 4 dwelling units and 13 residents. Camden remains the county's largest city. Economic development is a regional phenomenon, and as long as 56 cities continue arguing over matters such as new rail mass transit development, public school funding, parks development, and other planning issues that overlap municipal boundaries, overall planning, development, and revitalization efforts for the greater Camden area will be compromised. Federal and state government can encourage regional collaboration with a variety of carrots and sticks such as the Clean Air Act and ISTEA (Rusk, 1993).

Dreier (1997, pp. 14-15) discusses the major role played by Community Development Corporations (CDCs) and other nonprofit organizations in Boston's revitalization during the 1980s. Camden has developed several energetic and committed grassroots organizations in recent years. Examples are the Camden Churches Organization for People (CCOP), which has spearheaded several successful nonprofit housing ventures, and the Latin American Economic Development Association (LAEDA), which has enjoyed success with commercial revitalization in Camden's neighborhoods and the downtown area. These nonprofits also have been active in neighborhood planning. One group, the North Camden Association, produced a neighborhood plan that won an American Planning Association Award for innovation in citizen participation (Riordan, 1993). These nonprofits are, for the most part, not linked to either city or county political machines and are headed and staffed by eager, hardworking, idealistic visionaries who have decided to remain in Camden and make it a better place to live. Direct funding to these nonprofits by state, federal, and private sources, bypassing City Hall roadblocks, could be part of a solution.

## CONCLUSION

Up until the mid-1950s, Camden was a thriving industrial city. Since that time, the factories have either closed or left town, along with most of the private sector businesses. First, the white middle and working classes fled Camden, and since 1990 much of the black middle class has left the city as well. Since the 1970s, Camden's mayors have been ineffective in stabilizing the city's economic base and rallying citizen constituencies for change. Camden still has mid-sized medical centers and a branch of Rutgers University within its borders, and it has some 90,000 residents, a population that has been relatively stable since 1980. To date, however, these factors have not been enough to boost revitalization.

On the other hand, the city has locational advantages superior to cities of similar size in the greater region that have experienced moderate success in downtown revitalization (Wilmington, Delaware, and Allentown, Bethlehem, Wilkes-Barre, and Scranton, Pennsylvania). For Camden to mirror these cities' successes, strong local governmental leadership must come to the surface, comprehensive planning must be utilized, metropolitan cooperation must be heightened, and the various neighborhood-based nonprofit organizations committed to housing and commercial revitalization must be empowered. Although private business in partnership with state and federal government can help, the impetus for change must come from within the city itself. The present Empowerment Zone can be a framework for change but cannot substitute for the interventions mentioned above.

## NOTES

1. On March 8, 1997, U.S. Department of Housing and Urban Development secretary Andrew Cuomo praised Philadelphia's handling of its $79 million share of Empowerment Zone money but sharply criticized Camden's lack of progress. In Philadelphia, 400 new jobs were created in the Empowerment Zone, along with 44 businesses assisted and 3 existing businesses expanded. Philadelphia also had committed $67 million of the $79 million to loan programs, business relocations, and expansion efforts to retain jobs. Camden, on the other hand, had done nothing. On May 19, 1997, the Camden Empowerment Zone Corporation (CEZC) awarded its first planning contracts, in the amount of $28,000.

2. On May 13, 1997, Camden voters rejected Dr. Arnold Webster's bid for a second term. In a binding nonpartisan election for mayor, City Council President Milton Milan, a 34-year-old Puerto Rican businessman, won the post with 40% of the votes cast. Dr. Webster received only 26%, and two other African American candidates won another 26% between them. Milan has the support of state government, but it will take time before one can assess his performance in office.

On July 1, 1998, the state of New Jersey took control of the city of Camden's finances by installing a seven-member Oversight Board. Governor Christine Todd Whitman named six state officials, headed by Stephen Sasala and Milan, to this body, which must approve all capital and operating spending (Nedo, 1998, p. 1A). On November 12, 1998, the state of New Jersey took control of the Camden Police Department, citing inadequate neighborhood patrols, ineffective communications, and unacceptable delays in police response time. State Attorney General Peter G. Verniero appointed Camden County Prosecutor Lee A. Solomon as a monitor to oversee the department (Ott, Phillips, & Jennings, 1998, p. B1).

# 5

# CHICAGO
## Community Building on Chicago's West Side— North Lawndale, 1960-1997

Robert Giloth

## INTRODUCTION

In 1936, the primarily Eastern European neighborhood of North Lawndale on Chicago's west side represented neighborhood stability and
cohesion, delivering 24,000 votes for the presidential campaign of
Franklin D. Roosevelt, an incredible 97% of the ballots cast. He called it
the "best Democratic Ward in the country" (Lemann, 1991, p. 82). In
1986, the 5.5-square-mile neighborhood of North Lawndale had become—for a team of *Chicago Tribune* reporters—the emblem of failed
Great Society social policies and a neighborhood dominated by the
pathologies of the underclass, an "American Millstone" (*Chicago Tribune*, 1986).[1] Today, in another shift, after the community declined from
a peak population in 1960 of 120,000 to some 45,000 in the 1990s, a
10-theater Cineplex and middle-income townhouses are being built,
North Lawndale home values appreciated faster in 1995 than in all but

one Chicago neighborhood, and community organizers are again wor-
rying about real estate speculation.[2]

What happened in North Lawndale during the past 50 years illustrates that urban neighborhoods are neither homogeneous nor entirely
of their own making. Rather, they are subject to micro and macro
dynamics that spatially order and differentiate metropolitan landscapes
by class, race, and land use (Jackson, 1985; Massey & Denton, 1993;
Perin, 1977). North Lawndale changed from a largely white neighborhood to an exclusively African American neighborhood in less than a
decade, experiencing the chaotic interplay of the northern migration of
southern blacks, pent-up black housing demand, financial disinvestment in urban neighborhoods, suburban housing construction, and
white flight and "upward mobility" to outlying city neighborhoods and
the suburbs (Hirsch, 1983; Massey & Denton, 1993). A massive loss of
125,000 manufacturing jobs followed in and around the neighborhood
in the 1970s and 1980s, undermining the economic ladders upon which
past city immigrants had pulled themselves up. The outcome of this
process is a "green and institutional ghetto" of vacant land, abandoned
buildings, and fortress architecture (Vergara, 1995).

What distinguishes North Lawndale among many neighborhoods
that have experienced similar dynamics is that its plight occasioned a
series of ambitious and contrasting responses by community residents,
advocates, government, and business to build a community amid
change. These responses shaped, and have been shaped by, federal
urban and social policies. North Lawndale is layered in designations.
The effectiveness of these responses to neighborhood change, however,
were constrained and frequently undermined by the machine politics
and growth agenda of Chicago political elites, which, particularly under
Mayor Richard J. Daley (1955-1976), discouraged community action
independent of local ward politicians (Ferman, 1996; Squires, Bennett,
McCourt, & Nyden, 1987).

This chapter interprets North Lawndale's recent history, examining
a sample of community development strategies and organizations op-
erative in the neighborhood during the past three decades.[3] It explores
the fundamental question: How can we build enduring places that
sustain and nurture community in the midst of spatial and economic
change? The chapter begins with a history and socioeconomic overview
of North Lawndale, followed by discussions of civil rights and community organizing, community planning, community development corpo-

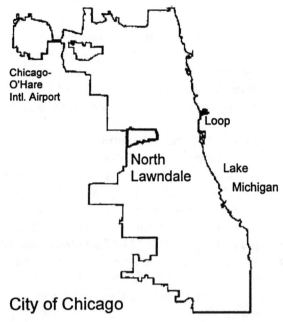

**Figure 5.1.** North Lawndale, Chicago
SOURCE: Created by City Design Center, University of Illinois at Chicago. Used with permission.

rations, faith-based development, market-driven housing, and community building. It concludes by offering several alternative interpretations of North Lawndale's community development history.

## A HISTORY OF NORTH LAWNDALE

In the nomenclature of Chicago School sociologists, North Lawndale was born a neighborhood of secondary settlement, of "workingmen's cottages": the next concentric zone out from the inner ring of industry, immigrant ports-of-entry, and warehouse districts surrounding central business districts (Burgess, 1925).[4] Located three miles from Chicago's downtown "Loop," North Lawndale is bounded by the Eisenhower Expressway (U.S. 290) on the north, Cermak Road on the south, Western Avenue on the east, and Cicero Boulevard on the west (see Figure 5.1). Industrial districts on the east and west surrounded a residential core.

North Lawndale's history is that of Chicago in the 19th and early 20th century: Its growth was spurred by the extension of train lines, the inner-city exodus after the Chicago Fire of 1871, and its function as "greenfields" for the construction of mass production factories and warehouses. By 1930, North Lawndale's population topped 112,000, more than 46% of whom were Russian Jews, and many of whom had moved from Maxwell Street on Chicago's near west side. Two-flat cottages and large corner rental properties defined its residential landscape.

This peak population masked the outmigration and neighborhood change that already were occurring in North Lawndale, leading to a population loss of 10,000 between 1930 and 1950 (Chicago Fact Book Consortium, 1995). The reason for this loss was typical of American city building: Upward social mobility translated into spatial mobility from the ethnic slums, despite the community roots evident in 40 synagogues and 3,500 volumes in Yiddish at the Douglas Library (Bowden & Kreinberg, 1981). At the same time, the small west side black community was expanding westward, and by 1950 the northern section of North Lawndale—adjacent to the East Garfield Park neighborhood—contained 11,000 African Americans.

What happened next has been described by urban geographers as a "blow out" (Harvey & Chatterjee, 1974). Whites left in a wholesale fashion in the 1950s, driven now not only by economic opportunities and by an expanding housing market but also by "racial" fears and exploitative real estate practices. Opposition to this change was small compared to what happened on the south side of Chicago (Hirsch, 1983). During the 1950s, North Lawndale's white population dropped from 87,000 to 13,000, while its black population increased from 13,000 to 113,000, mostly low-income migrants new to Chicago. North Lawndale's population peaked at 125,000 in 1960. What happened over the next three decades was equally dramatic: North Lawndale lost 30,000 residents per decade, reaching 47,296 according to the 1990 census. High fertility rates balanced this dramatic outmigration, producing an age structure in which 43% of the population was under the age of 21 in 1990 (Chicago Fact Book Consortium, 1995).

North Lawndale's population in 1990 combined the poor and working poor. North Lawndale's median family income of $15,190 was less than half of the median for the city, 44% of the families lived below the poverty line, and 27% of the labor force was unemployed in 1990. More than 16,000 residents over the age of 16 were not in the labor force. Not surprisingly, 27,731 people, or 59% of North Lawndale's population, received some form of public assistance in 1990, 12,397 being children (Chicago Fact Book Consortium, 1995).

The North Lawndale neighborhood ranked ninth highest in Chicago in terms of persistent poverty (London & Puntenney, 1993). North Lawndale's poverty rate of 48.4% was more than double Chicago's rate, which itself was 50% higher than the U.S. rate. Forty-six percent of the households with two or more children were female headed in 1990, and 29% of all births were to young women under the age of 20. Only 48.1% of North Lawndale's population over the age of 25 had a high school diploma. Thirteen percent of the fatalities in North Lawndale were related to alcohol and drugs.

The dramatic decrease in North Lawndale's population between 1960 and 1990 is related to the overall disinvestment in and demolition of its housing stock. More than half the housing stock has been lost since 1960, many of these units in multifamily buildings that anchored block corners, and 14% of the units were overcrowded in 1990. Seventy-five percent of the units are rental. The remaining owner-occupied units are owned predominantly by residents over 65, suggesting that many additional units will soon come on the market (Schubert, 1993). The average single home value in 1990 was $43,800, and banks made only 153 loans ($5.4 million) in 1991 (Chicago Rehab Network, 1993).

Loss of people and housing units occurred at the same time that North Lawndale's economic base was declining. International Harvester closed its plant the late 1960s, and by the mid-1970s, Sears had moved most of its headquarters staff to its new downtown skyscraper. Sears closed the rest of catalog distribution center in the mid-1980s, shortly after Western Electric closed its massive Hawthorne plant on the western border of North Lawndale. Although corridors of manufacturing and service firms remained, comprising 25,946 jobs concentrated in manufacturing, health care, and services, North Lawndale residents were underrepresented in their employment (Ducharme, 1997, pp. 15-16). What remained of North Lawndale's commercial thoroughfares burned or suffered fatal insurance and investment redlining after the riots following the assassination of Martin Luther King, Jr.

## CIVIL RIGHTS AND COMMUNITY ORGANIZING[5]

In 1966, Martin Luther King, Jr., rented an apartment at 1321 South Hamlin in which he planned to spend 3 days a week until the beachhead of the northern civil rights campaign produced victory. Chicago was chosen because of its community organizing against the discriminatory practices of school superintendent Benjamin Willis and because New York's Adam Clayton Powell had rebuffed Dr. King. King and his organizers called it the "End the Slums" campaign (Garrow, 1986; Lemann, 1991).

What the "End the Slums" campaign would entail eluded Dr. King and his organizers for months, in part because of their global understanding of the slum. In the words of Dr. King, "The slum is little more than a domestic colony which leaves its inhabitants dominated politically, exploited economically, [and] segregated and humiliated at every turn" (Garrow, 1986, p. 466). But what was to be done? They planned a three-phase campaign that would, at an incredibly fast pace, reach scale within 6 months, culminating in "massive actions" that would achieve "a direct confrontation" between "the power of the existing social order and the newly acquired power of the combined forces of good-will and the under-privileged" (Garrow, 1986, p. 457).

This ambitious, if vaguely defined, campaign got off the ground in fits and starts. Tenant organizing occurred, but there were embarrassing gaffs in the choice of targets.[6] Attendance at a mass rally at Soldier's Field ranged from 20,000 to 60,000, depending on who was counting, and 5,000 marched afterward with Dr. King to City Hall, where he taped a list of demands on the entrance door. Chief among these demands was the ending of housing discrimination by realtors and banks. Other demands included guaranteed family incomes, tenants' rights, desegregation of schools, and improved federal housing policies (Garrow, 1986).

What started as a social justice campaign anchored in North Lawndale ended in a standoff pitting Dr. King and Chicago's Coordinating Council of Community Organizations, led by Al Raby, against Mayor Richard J. Daley. The civil rights campaign took to the streets in white ethnic neighborhoods such as Gage Park, only to be met with rocks and jeering crowds. The penultimate march planned for Cicero, Illinois, was averted by a last-minute, negotiated agreement between King, Daley, and Chicago realtors to promote expanded open housing (Garrow, 1986). Paradoxically, the End the Slums campaign crystallized around open housing rather than the rebuilding of neighborhoods like North Lawndale. This victory reinforced the outmigration of moderate-income families from North Lawndale to other neighborhoods and to inner ring suburbs (Lemann, 1991).

Only a few blocks from where Dr. King rented his apartment in the winter of 1966 stood Presentation Parish, home to Monsignor Jack Egan, the most dogged supporter of community organizing and Saul Alinsky in the Chicago Archdiocese.[7] Egan started Operation Saturation, in which visiting seminarians and volunteers canvassed the blocks surrounding the parish every Saturday to engage residents in conversations about their lives. They were always on the lookout for "issues" and potential leaders. Jack MacNamara was one of those seminarians, and eventually one "brave parishioner" confided in him a "dirty secret" about North Lawndale's real estate market:

> Black buyers were forced to deal with one of five or six real estate speculators who bought properties and then sold them on contract. In the contract sale, interest was high and buyers had no equity in their property until they had made the last payment. A payment missed because of illness or job loss could put a family on the street. (Frisbee, 1991, p. 197)

This type of real estate transaction thrived because of the redlining of inner-city black communities by conventional banks and by the Federal Housing Administration (FHA) (Massey & Denton, 1993). Thus was born the Contract Buyer's League (CBL).

After MacNamara and a team of college students researched recent property sales in North Lawndale to uncover contract purchases, meetings with residents grew in size and frequency. Their first resident leader to fight back was Ruth Wells: She went with MacNamara and Egan to renegotiate her contract into a standard mortgage. Emboldened residents and organizers eventually formed a coalition of south side and west side homeowners, and in December, 1968, the CBL launched a rent strike against the contract sellers. Within 3 months, 600 contract buyers had placed $250,000 of rent in escrow funds. The CBL and its volunteer lawyers filed two lawsuits in U.S. District Court on behalf of the homeowners, arguing that their civil rights were violated by being forced into "involuntary financial servitude" by the contract sales (Douglas, 1970). When the court ruled that the cases be tried as class action suits, the CBL undertook a massive research effort on 475 contract sale properties. Meanwhile, contract sellers forced the eviction of 21 families (McPherson, 1972).

Court action dragged on for more than 10 years, resulting in eventual losses for the CBL in the appeals courts. During this period of community action, nevertheless, the CBL renegotiated 500 contract sales for a savings of $17 million, $13,000 per homeowner. By the early 1970s, however, the CBL had ceased community organizing; the courts and pro bono lawyers maintained its organizational momentum (Giloth, Meima, & Wright, 1984).

## PLANNING AND A COMMUNITY DEVELOPMENT CORPORATION

In December, 1966, the West Side Federation (WSF) discovered a "Negro removal" plan being secretly prepared by a major real estate developer, an architectural firm, and other civic and business interests to tear down a large part of North Lawndale for a "new town" that would protect the huge Sears complex (Houghteling, 1978).[8] The WSF was a coalition of west side churches, led by Rev. Shelvin Hall, that predated King's northern campaign and became a center of grassroots and civil rights organizing, including the takeover of 1321 South Homan with Dr. King (Anderson & Pickering, 1986). North Lawndale activists concluded that downtown and business interests promoted this plan for demolishing 190 blocks and for building 12,500 units of middle-and upper-income housing and a 45 acre golf course.[9]

The WSF transformed anger about this plan for neighborhood destruction into action at a June 6-7, 1967, conference titled "Today's Lawndale: Black Colony—Tomorrow's Lawndale: New City," sponsored by an ad hoc group of North Lawndale organizations and funded by the Maremont Foundation and the Community Renewal Society (*Chicago Tribune*, June 7, 1967, sec. 2A, p. 7). Fearing the possibility of a community uprising if the plan for neighborhood destruction came to light, the WSF investigated how it could undertake its own community plan. The organization brought in urban development advocates such as Claude Brown, Charles Abrams, Jane Jacobs, and Richard Hatcher to argue for the viability of urban neighborhoods and the need for community planning. The conference shaped initial ideas about how such a plan might be made. The strategy worked: By early 1968, the community had raised enough money to retain Greenleigh Associates of New York and Marcou, O'Leary, and Associates.

The plan they produced embodied a distinctive planning philosophy: "The basic redevelopment strategy might be described on the surface as 'Responsible Militancy.' The strategy is action oriented rather than reactionary" (Greenleigh Associates, 1968). The plan recommended forming an organization that would promote and represent communitywide consensus; consequently, in November, 1968, four organizations merged into the Lawndale People's Planning and Action Conference (LPPAC). Second, the plan recommended a specialized economic development organization to promote and sponsor North Lawndale's redevelopment. A new profit-making organization, originally named the North Lawndale Economic Development Corporation (NLEDC), was set up to undertake residential, commercial, and industrial development. It would be renamed Pyramidwest Development Corporation in 1974.

The notion of "prototyping" the plan became reality when LPPAC membership voted in 1968 to focus development in a 28-block area they called "Lawndale Center" (Greenleigh Associates, 1969; Houghteling, 1978). Prototyping included three components: a multiservice shopping center, a concentration of different types of affordable housing, and cultural/educational improvements—including a new high school. The LPPAC would pursue the building of the new George Collins High School in Douglas Park (completed in 1976), but much of the work required for the Lawndale Center would be that of the NLEDC.

Community residents organized the NLEDC and later Pyramidwest as for-profit corporations.[10] Pyramidwest's underlying development theory for building a new North Lawndale was to become a more self-sufficient economy. Cecil Butler, the NLEDC's president, stated in an internal memo that "The best we can expect to do is to become limited producers of goods that have some value to other areas and to be capable of generating enough income to support the inhabitants of this community" (Baron, 1975, p. 18).

This theory, over time, became more narrow: to undertake land and physical development on a significant scale, assembling and protecting land from other uses. Lewis Kreinberg, research director of the NLEDC, stated that "We are not preserving or rehabilitating . . . the area known as North Lawndale. We are creating . . . a community which has never been known in this West Side ghetto" (Casas & Colvin, 1974).[11]

Pursuing this strategy was made possible by NLEDC/Pyramidwest becoming part of the federal Office of Economic Opportunity's Special

Impact Program to support community development corporations (Parachini, 1980). Pyramidwest received more than $18 million of War on Poverty funds, as well as substantial amounts of project-specific funding from the Economic Development Administration (EDA) and the Department of Housing and Urban Development (HUD), to support its operations and to invest in promising community development opportunities.[12] There were five building blocks for Pyramidwest's plan:

1. The 16-acre Lawndale Plaza Shopping Center on Roosevelt Road, originally projected to contain 300,000 square feet of leased commercial space;
2. 1,600 units of affordable housing;
3. The redevelopment of the 60-acre Cal-West industrial park (the vacant International Harvester site in South Lawndale) for job-producing uses;
4. The Community Bank of Lawndale; and
5. Health care and communications ventures.

Pyramidwest's 1974 annual report, titled *It's About Time!*, ironically conveyed the challenges that dogged this community development corporation (CDC) into the 1990s (NLEDC, 1974, 1975). Although Pyramidwest secured most of the land for Lawndale Plaza by the late 1970s, only in 1998 was actual development anticipated by a variety of partners: 150,000 square feet of space was to include a 10-unit Cineplex theater, a supermarket, and other retail enterprises. Pyramidwest built senior housing and townhouses, and it became manager for defaulted HUD properties. In 1995, Pyramidwest received a $51 million FHA-insured Illinois Housing Development Authority bond to renovate 1,240 apartments in 100 buildings in 10 square blocks close to Douglas Boulevard (McRoberts, 1995). The Cal-West Industrial Park remained largely empty for years, a result of the market and Pyramidwest's reluctance to sell the land at a low price or to settle on businesses (like warehouses) that were not labor intensive. By the 1990s, Pyramidwest had sold the property because of delinquent taxes and public use requirements, in one case, tragically, for the expansion of Cook County Jail. The Community Bank of Lawndale opened in 1977, after several years of fighting with regulators and Sears, and in 1996 was spun off from Pyramidwest to become an independent bank with $40 million in assets (Bronstein, 1996).[13] Finally, Pyramidwest built and operated a 300-bed nursing facility in the Cal-West Industrial Park but failed in a prescient effort to develop a cable television franchise for Chicago.

The *Chicago Tribune's* 1986 muckraking series on North Lawndale, "The American Millstone" (see *Chicago Tribune*, 1986) assigns blame for the slowness of development in North Lawndale, and the community's continued demise, to Cecil Butler, the president of Pyramidwest, and the general antibusiness attitude of community control.[14] The paper asked why the "promise of 1968 was not kept." Butler's reply, at the time, was "I don't think it's realistic to measure our accomplishment by our plans and expectations, in the context of general economic conditions" (*Chicago Tribune*, 1986, p. 179). Neighborhood housing analysts have criticized Pyramidwest as making things worse by focusing on scale and housing supply issues rather than working to build on local assets and the existing housing market (Schubert, 1993). Indeed, a close reading of Pyramidwest's development projects shows the fatal combination of poor market conditions, collapsed urban policy after 1973, changing governmental regulations, bureaucracy, and high expectations (Houghteling, 1978).

## FAITH, MARKETS, AND COMMUNITY BUILDING

North Lawndale's economic and demographic fortunes continued to sink as the 1970s unfolded. The community's unraveling, combined with the slow implementation of Lawndale Center by Pyramidwest, encouraged new plans and development responses for North Lawndale during the 1980s and 1990s. The business community, uncomfortable with the community control rhetoric of the 1960s and 1970s, developed its own initiatives, including the Lawndale Business and Local Development Corporation to package loans and technical assistance for local businesses. It sponsored two major local planning exercises: Project 80 and a strategic development assessment of several commercial parcels by the Urban Land Institute (Project 80, 1982; Urban Land Institute, 1986).

Three specific innovations during this period contributed to a changed dynamic in North Lawndale that, by 1996, in the eyes of many North Lawndale observers, had created "pockets of hope" rather than an American Millstone. These innovations—Lawndale Christian Development Corporation, Homan Square, and the Steans Family Foundation—occurred in a changing neighborhood context. North Lawndale's location began to attract investment. On the east, the 560-acre Illinois Medical Center and its related development was moving westward, in part pushed along by the new United Center sports stadium—housing the Chicago Bulls and the 1996 Democratic Convention. From the south, the growing Mexican American community of Little Village was bursting at the seams, boasting the highest retail sales volumes after Chicago's famed Michigan Avenue.

## LAWNDALE CHRISTIAN DEVELOPMENT CORPORATION

In 1978, Pastor Wayne Gordon, a white evangelical missionary trained at Wheaton College, founded the nondenominational Lawndale Community Church. He came to North Lawndale as a teacher and coach. Gordon's mission represented a form of "Christian community development" that emphasized relocation to inner-city communities, reconciliation between races, and redistribution of resources, power, and hope to community residents (Gordon, 1995).[15]

Gordon's mission evolved rather quickly into one of service. A survey conducted by the Lawndale Community Church about why residents did not attend church revealed that they felt inadequate to participate and frequently ripped off by churches (Williams & Bakama, 1992). Gordon wanted to transform these perceptions. He developed an approach called the three Ps—people (listening), prayer, and partnerships (Gordon, 1995, pp. 64-67). In addition to providing space for residents to gather, the church built a gym and operates a clothing and food pantry and emergency housing for the homeless.

The church founded a number of programs to meet community needs. By 1996, the Lawndale Christian Health Center, formed in 1984, had 70,000 patient visits, 120 employees, and 29 doctors. The Lawndale College Opportunity Program provided counseling, academics, and funding for young people in the neighborhood to go to college. In 1987, the church founded the Lawndale Christian Development Corporation

(LCDC) to implement housing, economic development, and education programs. The LCDC has a staff of 13, an annual operating budget of $550,000, and an 11-person board consisting of eight community residents, two doctors from the health center, and a state representative.

The LCDC's development strategy is to rehabilitate buildings with two to four flats in the 40 square blocks surrounding the church, creating home ownership opportunities for existing residents. The LCDC has developed more than 20 lease-to-purchase units and has undertaken moderate rehabilitation of a 22-unit building and renovation of a 48-unit multifamily building. Recently, the LCDC received a HUD Hope 3 grant to purchase and rehabilitate 18 buildings with one to four units each. The LCDC has partnered with the Neighborhood Housing Services (NHS) since 1993 to make loans available to homeowners, lending $3 million between 1993 and 1996. To support these efforts, the city of Chicago has designated the area in which the LCDC concentrates as a SNAP (Strategic Neighborhood Action Planning) area for coordinated city investments of $3 million.

The LCDC has invested in economic development as well. It has renovated four commercial properties on Ogden Avenue, has launched three for-profit businesses, and is working with other churches and agencies to develop job training and placement programs in construction and for local industries. The LCDC plans to build a $3.5 million day care facility that will serve 250 children and employ 50 people in 1999, in conjunction with the Carol Robertson Center, the Illinois Facilities Fund, and the Empowerment Zone.

## HOMAN SQUARE[16]

North Lawndale's history has been fatefully intertwined with that of Sears Roebuck and Company. In 1973, Sears moved 7,000 employees from North Lawndale to its new headquarters in the downtown Sears Tower; in 1987, it moved its remaining 3,000 employees from its catalog operation in North Lawndale. What it left behind were four buildings, containing four million square feet in the heart of North Lawndale.

For several years, Sears marketed these properties for commercial uses. There were no takers. In 1988, the CEO of Sears, Edward J. Brennan, teamed up with Charles Shaw and Company, a major real estate developer of residential projects and downtown office buildings. During the next 3 years, they put together a plan for tearing down much of the old Sears plant, building 600 units of new housing, and marketing the headquarters building for commercial and public uses.[17] What was strikingly different about this plan was its mixture of housing types and affordability, reaching families with incomes as low as $15,000 and as high as $80,000. Shaw believed that North Lawndale was marketable and that mixed income development was the only path to sustainable neighborhood development.

Homan Square, to be built in three phases by the year 2000, already contains several hundred occupied units. Shaw spent 2 years cultivating support in the community for a planned development zoning approval. In particular, he had to ward off community fears of gentrification and displacement. Substantial public subsidies in the form of low-income housing tax credits, municipal housing finance, tax increment financing, and infrastructure have complemented Sears's investment in making the buildings good long-term investments. Prices of the homes for sale range from $85,000 to $175,000. Although most of the early buyers came from within the community, increasingly Homan Square is attracting African American in-movers from other city neighborhoods and even from the suburbs.

## THE STEANS FAMILY FOUNDATION

By the early and mid-1990s, a number of new and promising, albeit disconnected, community development activities were taking place in North Lawndale. These ranged from the LCDC and Homan Square to block clubs of old-time homeowners such as the "Slumbusters" on Flournoy Avenue, indigenous rehabbers, the Mount Sinai Hospital expansion, the Mid-America Leadership Institute, the Chicago Community Trust's Children and Family Initiative, and the sponsorship of Habitat for Humanity housing by Presentation Parish. Part of North Lawndale was included in Chicago's 1994 Empowerment Zone designation, and the city of Chicago applied for a $20 million Homeownership Zone for the 50-block area just south of Homan Square. A new Walgreen's drugstore was built at the corner of Roosevelt and Kedzie, and Lou Malnati's Pizza Restaurant opened on Ogden Avenue in a building renovated by the LCDC. North Lawndale also participated in a promising "reverse commuting" demonstration by Suburban Job Link that links community residents to jobs surrounding O'Hare Airport. Although the LPPAC had died by the 1990s after trading its planning and organizing role for human service delivery, Pyramidwest—in its various forms—remains a key development actor, if for no other reason than it owns strategic development parcels.

North Lawndale remained a community to be built, despite its "pockets of hope" and newly renewed economic location. What it needed was a process to knit together the activities, leaders, and organizations of North Lawndale—something similar to what was called for in the 1968 Greenleigh Plan's recommendation for the formation of the LPPAC. Such a convening force fortuitously appeared in North Lawndale in 1994 when the newly formed Steans Family Foundation made a long-term commitment to community building in North Lawndale. The foundation consisted of Harrison Steans, a major financial investor in 1st Chicago NBD Bank, and his three daughters. This was to be hands-on philanthropy for family members and staff.[18]

Several assumptions guided the Steans Family Foundation as it entered North Lawndale. First, its involvement "reflect[ed] a belief that people are embedded in families, social networks and institutions, and communities" (Brown, Pickens, & Mollard, 1996a, p. 1). Second, it believed that flexible, noncategorical dollars targeted to a specific geographic neighborhood could improve outcomes for children and families. Third, it believed that "healthy communities have multiple components and that . . . efforts must be mounted in each of these areas and linkages created between them" (Brown, Pickens, & Mollard, 1996b, p. 1). These assumptions are shared with other comprehensive community initiatives (CCIs) around the country (Connell, Kubisch, Schorr, & Weiss, 1995).

Steans chose North Lawndale in part because of the relationships it already had with local leaders such as Pastor Wayne Gordon. Steans decided to focus on investing in projects, hoping to build an overarching plan upon the experience of pursuing opportunities and organic evolution. Several principles guided its actions and investments from the outset: participation and ownership by residents, building individual and organizational capacity, promoting indigenous leadership, strengthening networks and connections among individuals and organizations, and strengthening connections to outside resources. Steans played a number of interrelated roles as it invested in North Lawndale: grantmaker, convener, broker, capacity expander, incubator of new ideas, and community advocate (Brown et al., 1996a, pp. 3-6).

Operationally, Steans has invested $1 million during 1997 in North Lawndale as part of a 5-year renewable commitment. The foundation's current assets are $25 million, and it pays out $1.5 million annually, 95% of which goes to North Lawndale. It hopes to leverage two additional dollars for every one it invests in education and youth development, housing, economic development and employment, health and human services, and quality of life. Its strategies include building a North Lawndale Learning Community of 10 elementary schools and Manley High School to improve achievement; program-related investments to the Community Bank of Lawndale to enhance small business lending; building a Family Resource Center; funding several groups to work with CANDO, a citywide coalition and technical assistance provider, to complete a feasibility analysis for a small shopping center; a small grants program for block-level community improvements; and incentives for community organizing groups to develop leadership and a common framework for action (Brown et al., 1996b).[19]

Steans is especially interested in bringing a full-service Neighborhood Housing Services (NHS) office to North Lawndale to provide comprehensive home ownership services. In contrast to major housing supply interventions such as Homan Square or Pyramidwest, an NHS would build on the home ownership market that currently exists, removing barriers and adding incentives to make it work better. The overall purpose would be to change the way small investors make decisions about North Lawndale (Schubert, 1993). In the same fashion, Steans recently commissioned a study of economic development and job-related possibilities in North Lawndale (Ducharme, 1997).

## NORTH LAWNDALE'S PAST AND FUTURE

Driving around North Lawndale in 1998 does not obviously nor dramatically reveal pockets of hope, the imminent building of a shopping center, or community building. What one sees most is vacant land. An associate at Charles Shaw and Company nevertheless described North Lawndale's collective pace of development as "being on second base."[20]

The city of Chicago has bet on North Lawndale's rebirth, investing close to $50 million in North Lawndale in the last 5 years and making it one of its primary redevelopment areas.[21]

The North Lawndale story challenges urban and neighborhood development theory. Revitalizing inner-city communities requires the right balance of favorable external factors and internal capacities for neighborhood organization and development (Keating & Smith, 1996). Community organizing, planning, or development alone cannot easily overcome the forces of disinvestment in neighborhoods experiencing dramatic change.

The North Lawndale experience highlights two central questions about neighborhood revitalization. First, it gives new life to a framework posed by urban economist Anthony Downs several decades ago about the "life cycle" of neighborhoods and whether some older neighborhoods ought to be "triaged" until they empty of population and buildings, eventually becoming economically valuable again (Downs, 1981). A recent adaptation of this theory has called for land banking of inner-city land for future development (Rybczynski, 1995). North Lawndale's massive loss of population, housing, and density was not sufficient by itself to spur redevelopment without the renewed importance of its economic location and an infrastructure of private, public, nonprofit, and community stakeholders. Moreover, North Lawndale's demise was not a natural process but instead the outcome of public, private, and individual decisions.

The second challenge concerns the techniques and practices of comprehensive community development. The 1968-1969 Greenleigh community plan for North Lawndale recommended a comprehensive approach, and the formation of Pyramidwest and the LPPAC recognized the importance of pursuing multiple dimensions of community development. The emphasis was on scale and protecting the land from bad uses like high-rise public housing or urban renewal. Yet despite past and continuing accomplishments, these organizations failed to sustain concerted community building. At the same time, serious rifts grew between a "community controlled" development process and the roles and resources of key North Lawndale stakeholders such as Sears, Ryerson Steel, Mount Sinai Hospital, and the city of Chicago.

The turn to small-scale and frequently faith-based development, private development such as Homan Square, and the community building investments of the Steans Family Foundation offered an alternative. It attracted multiple and diverse partners, worked with people one by one and with neighborhoods block by block, and attempted to build neighborhood markets and assets such as home ownership. These approaches eschewed central planning and organization, opting instead for multiple centers of innovation and action, yet at some point the community must build additional community capacity for development.

By 1998, North Lawndale had come full circle in community building approaches. Today, many North Lawndale activists see the most promise in the congregation-based community organizing of the Metropolitan Sponsors, an affiliate of Saul Alinsky's Industrial Areas Foundation.[22] Presentation Parish, St. Agatha's, the Lawndale Community Church, and several others have become members, many of them the same churches that joined Dr. King, the Contract Buyer's League, and community planning in Lawndale. Leaders in these churches believe they need a more powerful voice and coalition if they are to grapple with the many internal and external challenges still facing North Lawndale.

The North Lawndale story is about the diaspora of African Americans to northern cities, among neighborhoods, and to inner ring suburbs. It is about the rigors, obstacles, exploitation, and false paths of this process and, in particular, the difficulties of low- and moderate-income people building and sustaining community. The history of community development in North Lawndale during these decades offers an array of community organizing, planning, and development efforts—with many victorious moments in a context of continuing neighborhood disinvestment and failed local and federal urban policies. What is redemptive about the North Lawndale experience, however, is that—by some accounts at least—a new generation of community building activities matched with the economic rediscovery of North Lawndale may produce the foundation for a neighborhood future.

## NOTES

1. The American Millstone series by the *Chicago Tribune* is representative of a decade-long process of socially constructing the meaning of "underclass" to mean bad, undeserving people (Gans, 1995).

2. The *Chicago Realtor* reported that North Lawndale had the second largest increase in average property values from 1994 to 1995—$44,500 to $78,000, or 75%.

3. For a more comprehensive examination of the history and revitalization of North Lawndale in comparison with the Englewood neighborhood, see Zielenbach (1998).

4. The boundaries of North Lawndale offer a convenience for discussing neighborhood revitalization on Chicago's west side, a much larger unit of analysis upon which many community organizing and development strategies have been premised (see Figure 5.1). Chicago School sociologists inventoried 77 community areas in the 1930s, among which is North Lawndale—Community Area 29.

5. The Greater Lawndale Conservation Commission operated from 1954 to the mid-1960s. It also saw itself as a link between Lawndale and citywide social agencies such as the Welfare Council. This section on neighborhood change and community voices does not consider the story of west side street gangs like the Conservative Vice Lords, which, by the late 1960s, had developed a distinctive, if short-lived, approach to community development (Dawley, 1972).

6. Dr. King participated in a community action that seized a South Homan street building in order to repair it. It turned out that the owner was an ailing 81-year-old. After several failed attempts to renovate the building, it was demolished in 1967 (Anderson & Pickering, 1986). The broader tenant organizing strategy also had questionable long-term impacts. A rent strike in 11 buildings housing 1,000 residents was turned into partial success in 1967 when 6 of the buildings were renovated by the Lawndale Freedom Movement/Kate Maremont Foundation with federal help. Unfortunately, financial, development, and bureaucratic problems soon drove this nonprofit charity out of business (Bowley, 1978).

7. Until the arrival of Archbishop Cody in 1965 from New Orleans, Egan had headed the Council on Urban Affairs of the Archdiocese. He was soon exiled to work out of Presentation Parish, too much a threat to be allowed to remain close to the center of power. His exile was extended farther, to Notre Dame University, in the 1970s, but he returned to Chicago under Archbishop Bernadin.

8. Community activists discovered this plan and prevented its formal release. The plan, developed by a committee of the Metropolitan Housing and Planning Council (MHPC), combined urban renewal of Lawndale with a proposed relocation strategy of blacks to the suburban town of Weston, Illinois, the site of the federal Argonne Labs (Metropolitan Housing and Planning Council, 1967). The MHPC pressured the city to target its Model Cities and Urban Renewal funds for this project, but the city was reluctant (*Chicago Tribune*, June 7, 1967, sec. 2A, p. 7). One observer concluded that the plan never had the political backing to move forward (Baron, 1974). (This account is based on interviews with Lewis Kreinberg of the Jewish Council on Urban Affairs, October 31, 1996, and January 2, 1997.)

9. Interview with Lewis Kreinberg, January 2, 1997.

10. Baron notes that a number of the initial organizers of Pyramidwest were lay leaders of Presentation Parish (Baron, 1974). Pyramidwest was formed as a Delaware corporation with three classes of stock: Class A common voting shares sold to community residents, Class B nonvoting stock held by Illinois Trust for the purpose of supporting the LPPAC, and Class C nonvoting stock for private offerings. Pyramidwest formed several subsidiaries and related organizations in the 1970s: Pyramidwest Realty and Management and California Gardens (a nursing home facility), a bank holding company for the Community Bank of Lawndale, a number of limited partnerships for housing development, and a nonsubsidiary nonprofit—the Local Redevelopment Authority of Lawndale, Inc.—to receive federal funds for economic development.

11. The Medill School received a grant to help the LPPAC put out a community newspaper, initially *The Black Truth*, later renamed *The Drum* (1965-1975). Casas and Colvin produced many of the articles for the paper that are assembled separately in this document. Ironically, it uses a format similar to what the *Chicago Tribune* reports used for their "American Millstone" series 10 years later, but with a hopeful spin on the challenges North Lawndale faced.

12. Pyramidwest withdrew from the Special Impact Program in 1976 over a dispute about the use of monies for the Cal-West Industrial site (interview with Robert Brandwein, October, 1996).

13. The bank made very few loans in North Lawndale and was under a "cease and desist" order from the Federal Deposit Insurance Corporation (FDIC) in the early 1990s. Some community development analysts are skeptical about the bank's impact (Schubert, 1993).

14. Cecil Butler graduated from Northwestern Law School in 1966 and arrived in North Lawndale soon thereafter. He was initially executive director of the LPPAC and became the first president of the NLEDC and Pyramidwest.

15. As part of this mission, Gordon stepped aside as pastor in 1995 so that an African American, Carey Casey, could take the lead as pastor. Gordon also stepped down from several related boards and took a sabbatical devoted to the Christian Community Development Association. These steps, in his mind, represented "giving power away" (Gordon, 1995, pp. 179-193).

16. The discussion of Homan Square greatly benefited from discussions with Mark Angelini, vice president of Charles Shaw and Company, December 22, 1996.

17. Interview with Angelini, December 22, 1996.

18. Interview with Greg Darnieder, Executive Director, Steans Family Foundation, December 22, 1996. Darnieder, a graduate of Wheaton College like Wayne Gordon and David Doig, previously operated youth programs in the Cabrini Green public housing projects.

19. These groups include a chapter of ACORN, Interfaith Action, and the new Industrial Areas Foundation coalition—Metropolitan Sponsors.

20. Interview with Mark Angelini, Vice President of Charles Shaw and Company, December 22, 1996. Amy Lozano, a planner with the city's Department of Planning and Development, was less sanguine, although she reported that Homan Square has stimulated the interest of other developers of moderate- and middle-income housing (interview with Amy Lozano, Department of Planning and Development, February 12, 1997).

21. Interview with David Doig, Assistant Commissioner with the Department of Housing, December 20, 1996. Doig was formerly Executive Director of the Lawndale Christian Development Corporation and also graduated from Wheaton College. The Neighborhood Capital Budget Group estimates that North Lawndale received $21,933,483 in completed infrastructure projects between 1990 and 1996, with $8,173,400 planned for the period 1997-2001 (Neighborhood Capital Budget Group, 1996). The city of Chicago is also coordinating the development of a physical plan for the portion of North Lawndale north of Homan Square that focuses on streetscapes, parks, schools, and vacant lots.

22. Interview with Richard Townsell, Executive Director of the Lawndale Christian Development Corporation, March 2, 1997. See also Ducharme (1997).

# CHAPTER 6

# CLEVELAND

*The Hough and Central*
*Neighborhoods—Empowerment*
*Zones and Other Urban Policies*

Norman Krumholz

## CLEVELAND: AN OVERVIEW

By the early years of the 20th century, Cleveland emerged as one of the world's manufacturing centers. Iron and steel mills, foundries, and automobile, clothing, and chemical factories gave substance to a grow-ing population that rose from 381,768 in 1900 to 560,663 in 1910. Immi-gration from Southern and Eastern Europe provided much of the increase. Workers tended to cluster in dozens of ethnic enclaves, while a trickle of more affluent Clevelanders moved further out. Almost all—rich and poor—lived in the same city, with the same institutions and political leadership.

World War I put an end to large-scale European immigration, but southern blacks were recruited by Cleveland industries. In a single

AUTHOR'S NOTE: This chapter benefits greatly from the comments and camaraderie of Bob Brown, Dennis Keating, Michael LaRiccia, Bill Resseger, and Jordan Yin.

decade, between 1910 and 1920, the city's black population increased 308 percent—from 8,448 to 34,451. Subject to growing racial discrimination, most of the new arrivals settled in the Central neighborhood, the site of Cleveland's first black ghetto (Kusmer, 1978, 1995).

By the 1930s, there were great differences between the city and its suburbs—differences in race, nativity, employment, and wealth. Population was increasing on the periphery and decreasing at the center. Cleveland's Eastern European ethnic neighborhoods were dispersing while the black community was growing and becoming ever more concentrated. As a local statistician observed in 1931, the city was "decaying at the core" (Green, 1931).

The pace of suburbanization increased after World War II as political and business leaders struggled to define Cleveland's problems. Like "growth coalitions" in other cities, they chose to strongly support and implement two federal programs—urban renewal and the interstate highway system—that dramatically and permanently changed the face of the city (Mollenkopf, 1983).

Cleveland's urban renewal program encompassed 6,060 acres (one-eighth of the city's entire land area) in seven city neighborhoods, all on the east side (Keating, Krumholz, & Metzger, 1989). It was the largest program in the nation and displaced thousands of mostly black, low-income Clevelanders from the Central, Mount Pleasant, Glenville, and Hough neighborhoods. But only in the downtown Erieview project was urban renewal successful in attracting substantial new private investment. Elsewhere, cleared and vacant land blighted east side neighborhoods. As late as 1976, almost 30% of the city's urban renewal land was vacant and unsold. Meanwhile, the construction of interstate highways 77, 71, and 90 displaced 19,000 more Clevelanders by 1975, also resulting in a significant loss to the city of both income and property taxes (Swanstrom, 1985).

Although Cleveland had been losing population since 1950, the exodus accelerated dramatically after 1960. Between 1960 and 1970, the city lost more than 125,000 residents, including 25% of its families with incomes over the Standard Metropolitan Statistical Area (SMSA) median. By the early 1970s, some 20,000 residents were leaving the city each year, and by 1990, Cleveland had lost about 45% of its 1950 population. During those 40 years, the city's population dropped from 914,000 to 505,000. The exodus emptied many neighborhoods of housing and of commercial and industrial buildings, and resulted in widespread abandonment of property, demolition of thousands of buildings, and vacant lots. To deal with the many thousands of these abandoned properties, Cleveland successfully lobbied to change the state law in 1975 to establish a municipal land bank to assemble, clear title to, and resell these parcels. By 1995, the city's land bank owned large tracts of land on the city's east side: 16% of all property in Hough, 13% in Fairfax, and 6% in Glenville.

Cleveland's jobs joined the exodus: Between 1958 and 1977, Cleveland lost 130,000 jobs, while the suburbs of Cuyahoga County gained almost 210,000. The job loss continued through the 1990s, with the city losing 35,000 additional jobs between 1990 and 1993 (Zeller, 1993). Manufacturing jobs were particularly hard hit. Between 1970 and 1985, the city lost more than 86,000 manufacturing jobs. Whereas in 1970, manufacturing provided 47% of all jobs in greater Cleveland, by 1987 manufacturing provided only 27%.

Problems of poverty and racial segregation in the city continued to escalate. Nearly three-quarters of the county's poor (about 215,000 people) lived in Cleveland, where the 1994 poverty rate was estimated at 42%. City neighborhoods, plagued by drugs and drug-related crime, continued to erode, and more than 17,000 Cleveland residents lived in what one federal official described as "the second worst public housing" in the nation (Jordan, 1989). Meanwhile, Cleveland had the second highest index of racial segregation of any city in the United States, a level of segregation so severe that the city was considered "hyper-segregated" (Massey & Denton, 1993).

Despite these problems, Cleveland in the 1980s and 1990s witnessed a remarkable downtown building boom. British Petroleum (formerly Standard Oil of Ohio), Ohio Bell, Medical Mutual, and Eaton Corporation all built new downtown office headquarters. Society Bank (now Key Corp.) built the city's (and the state's) tallest office building at 56 stories. These new office buildings helped hold downtown employment steady at about 125,000 jobs, although they also increased vacancy rates in older office buildings. In 1996, downtown vacancies stood at 19%.

Downtown also began to emerge as an entertainment district. An ambitious restoration project restored the Ohio, State, and Palace theaters on Playhouse Square for ballet, opera, and other entertainment events. The city's industrial flats were converted to popular restaurants, shops, and nightclubs.

In 1994, the $450 million Gateway project provided a new stadium for the Cleveland Indians baseball team and a new arena for the Cavaliers of the National Basketball Association. Two other attractions—a $94 million Rock and Roll Hall of Fame and a $56 million Great Lakes Museum of Science and Industry—were completed on the city's Lake Erie shoreline. Most of these projects were public-private partnerships, with by far the largest portion of the funding coming from public subsidies. These tax subsidies often diverted revenues from the deficit-ridden Cleveland school district, which, in 1995, was put into state receivership. The Cleveland Teachers' Union maintained that from 1990 to 1995 at least $35 million had been lost to the schools through tax abatement (Stephens, 1997). In 1995, the troubled Cleveland public schools graduated only 32% of all students who had enrolled in the 9th grade 4 years earlier; 68% dropped out or otherwise disappeared. Proponents of the downtown projects proclaimed that tax revenues lost to abatements would be more than recovered by new jobs and taxes for Cleveland residents. To date, that is an unfulfilled promise.

Cleveland in 1997 was a far different city from what it once was. Once the nation's sixth largest city, it was now twenty-sixth in size; once home to 75% of all residents in Cuyahoga County, it was now home to only 36%; once the home of affluent managers of Cleveland industry and a powerful working class, it was now home to most of the poor and black families in the region. Cleveland's families earned just 54% of the income of suburban families. Despite robust development downtown, the city's neighborhoods and public institutions continued to languish. As a reporter for the *Washington Post* observed, "The new Cleveland is corporate headquarters, service and professional jobs, and downtown construction. The old Cleveland is neighborhoods struggling against decay, double-digit unemployment, racial tension, poverty and long-suffering schools" (Sinzinger, 1986).

## HOUGH

The Hough (pronounced huff) neighborhood is a 2-square-mile area on Cleveland's east side bounded by Euclid and Superior Avenues and East 55th and East 105th Streets. In the late 1880s, Hough was a fashionable residential neighborhood characterized by large single-family houses. It was still a predominantly white, working-class neighborhood as late as 1950. Between 1950 and 1960, however, Hough underwent a

**Table 6.1**   Hough: Neighborhood Indicators by Census Tract, 1950-1990

|  | 1950 | 1960 | 1970 | 1980 | 1990 |
|---|---|---|---|---|---|
| Total population | 65,424 | 71,575 | 45,417 | 25,330 | 19,715 |
| Total housing units | 23,118 | 22,954 | 17,441 | 11,636 | 9,384 |
| Number of owner-occupied units | 4,281 | 3,802 | 1,480 | 2,208 | 1,837 |
| % owner-occupied units | 18.52 | 16.56 | 8.49 | 18.98 | 19.58 |
| Median housing value ($) | 7,949 | 12,286 | 11,975 | 15,320 | 21,694 |
| Number of one-dwelling units | 2,967 | 4,337 | 2,817 | 2,502 | 2,860 |
| % one-dwelling units | 12.83 | 18.89 | 16.15 | 21.50 | 30.48 |
| Number of residents 5 years + | 49,475 | 13,303 | 17,799 | 12,866 | 10,407 |
| % residents 5 years + | 75.62 | 18.59 | 39.19 | 50.79 | 52.79 |
| Median contract rent ($) | 42 | 74 | 77 | 100 | 165 |
| Median annual family income ($) | 2,866 | 4,626 | 4,906 | 8,824 | 10,990 |
| Median contract rent as % of median annual family income | 17.59 | 19.20 | 18.83 | 13.60 | 18.02 |
| Median family income as % of county median | 90.90 | 66.63 | 52.01 | 39.98 | 30.74 |
| % families below poverty level | —[a] | —[a] | 39.37 | 38.69 | 50.62 |
| % households receiving public assistance income | —[a] | —[a] | 42.19[b] | 37.49 | 47.81 |
| African American population | 2,562 | 52,710 | 42,210 | 24,305 | 19,220 |
| % African American | 3.92 | 73.64 | 92.94 | 95.95 | 97.49 |
| Number of families | 17,730 | 15,589 | 9,416 | 5,750 | 4,173 |
| Number of female-headed families | —[a] | —[a] | 3,600 | 2,718 | 2,575 |
| % female-headed families | —[a] | —[a] | 38.23 | 47.27 | 61.71 |
| % high school graduate or higher | 36.97 | 25.97 | 25.83 | 37.39 | 43.92 |
| % civilian labor force unemployed | 6.25 | 14.64 | 12.05 | 17.78 | 30.53 |

SOURCE: U.S. Census data.
Note: Data are presented for census tracts 1121 (L-1), 1122 (L-2), 1123 (L-3), 1124 (L-4), 1125 (L-5), 1126 (L-6), 1127 (L-7), 1128 (L-8), 1186 (R-6), and 1189 (R-9).
a. Data not reported by the census for this category in this period.
b. For 1970, the percentage of households receiving public assistance is reported as the percentage of families receiving public assistance (not households).

complete racial and class reversal, from 3.9% black in 1950 to 73.6% black in 1960, and from a median family income that was 91% of the county median to one that was only 67% of the median (see Table 6.1).

The rapid racial transition in Hough was begun by urban renewal projects in the abutting Central ghetto. These projects demolished much more housing than they built and contributed to severe overcrowding and deterioration on the entire east side of Cleveland, worsening an already bad housing situation for Cleveland's black families.

The St. Vincent project in Central illustrates the problem. The project was planned for predominantly residential use of 118.1 acres when the contract with the federal government was originally signed in 1955. As was later dryly reported, there was a premature exodus of families from this area. No record indicates where the premature exodus situated, but all evidence shows that the Hough area was most affected by the premature relocations; the exodus totaled approximately 1,200 families (Witzke, 1967). When no builders came forward to put up the new housing in the plan, the city Department of Urban Renewal took action in 1963 to change the re-use of the land from mostly residential to completely institutional. When it was all over, it was estimated that 1,780 families had originally resided in the project area. All left. No housing was built for the persons displaced, nor was relocation assistance provided. Relocation authorities had records of only 528 displaced families and admitted that the majority of black families relocated to areas heavily segregated, more than 90% black. Most of the land in the St. Vincent project eventually was sold to Cuyahoga Community College, St. Vincent's Hospital, and other civic institutions. Because most of these institutions were tax exempt, the St. Vincent project resulted in a net loss of assessed valuation for the city, from $2.3 million in 1953 to $2.1 million in 1972 (Olson, 1973).

As Central area blacks, fleeing the urban renewal bulldozer, scrambled into Hough for replacement housing, they were met by zoning restrictions to keep out low-cost housing. Neighborhood improvement associations such as the Hough Area Council were formed in the 1950s to campaign against signs of decay and generally to discourage black entry into the neighborhood.

When, in the late 1950s, the Hough neighborhood finally gave way to pressure by blacks on housing, Realtors practiced block-busting tactics and created panic selling at discount prices by white homeowners. Homes were then rented or sold at premium prices, reflecting the high demand for housing by blacks. As the white population pulled out, landlords converted many large, single-family homes into multifamily tenements and rooming houses in disregard of city housing codes. City records from 1956, for example, show that the owner of two houses and a barn converted the three buildings into 33 dwelling units (*Civic Vision 2000 Citywide Plan*, 1991). Conditions in Hough worsened as poverty and disinvestment increased.

Hough, the *Cleveland Press* reported in a series in February, 1965, was "in crisis." Racial violence erupted on the night of July 18, 1966, marking the start of the devastating Hough Riots, which lasted 7 days, cost four lives and millions of dollars in damage, and convinced many—white and black—that they had no future in Hough. Population and economic investment plummeted. Twenty-six thousand residents left Hough during the 1960s, and overall, the population dropped from 65,000 in 1950 to under 20,000 in 1990. By 1990, median family income in Hough was $10,990, and the median housing sale was $21,600.

In response to the Hough Riots of 1966, millions of dollars of federal money began to pour into the neighborhood for housing, job training, youth activities, health care, Model Cities, and other programs. One of the most significant organizations to emerge at this time was the Hough Area Development Corporation (HADC), established in 1967 with initial funding of $1.6 million from the Office of Economic Opportunity. The HADC was one of the first Community Development Corporations (CDCs) in the nation founded under the Economic Opportunity Act. The HADC tapped federal housing subsidies and local foundation support to build more than 300 new and rehabilitated housing units in Hough.

Commercial development ventures by the HADC often were less successful. An indoor shopping mall called Martin Luther King Plaza was put into receivership in 1982 and failed to produce any hoped-for economic development spin-offs. Community Products, Inc., an injection-mold rubber company organized to provide employment opportunities for Hough residents, failed to adjust to a shrinking auto parts market and went out of business in 1979. Several other HADC business enterprises also failed.

In 1984, the Reagan administration cut off the HADC's funding. The organization turned over its remaining assets to the city and went out of business. The HADC built housing but overall, its contribution to the physical stabilization of the neighborhood generally is seen as minimal.

Another organization, Famicos, emerged in 1969, a few years after the riots. Begun by a charismatic nun named Sister Henrietta, Famicos devoted itself to housing the poor (Wolff, 1990). In the process, Famicos developed an innovative, no-frills program to acquire, rehabilitate, and lease deteriorated housing to very-low-income families. Famicos's "lease-purchase" program involves multiple participants. Funds for acquisition and rehabilitation of housing come from the city through the Community Development Block Grant (CDBG), from local and national foundations, and from corporate investors seeking low-income housing tax credits. The housing is then renovated to code standards and leased to low-income families for monthly rents of about $225. The families are trained in the rudiments of housing maintenance and contribute some of the interior renovation work themselves through "sweat equity" to reduce rehab costs. Their monthly payments are applied against the outstanding mortgage, and the tenants have an option to buy the home after 15 years for the remaining balance. Through the lease-purchase program, tenants might become homeowners, whereas otherwise they might never have had a chance. In 1995, the "average" tenant was a minority single woman with three children, living on less than $8,000 a year (Krumholz, 1996).

The Famicos lease-purchase program not only served Famicos well but by 1995, the program also was in use by the 14 neighborhood-based nonprofit housing corporations that made up the Cleveland Housing Network, a housing umbrella organization. More than 1,500 homes have been rehabilitated citywide under the program, and Famicos continues to be heavily involved in the production of low-cost housing. Famicos also served as the nonprofit sponsor of Lexington Village, the first new large-scale housing development built in Hough in 40 years.

Lexington Village is a garden apartment rental complex located at East 79th Street and Hough Avenue, the flash point of the Hough Riots and the center of the bombed out and abandoned land in the neighborhood. The first phase of the project, 277 units, was completed in 1985. Subsequent phases in the planning stage include 432 more units. Rents start at $285 a month and, in keeping with the original terms of the project, 20% of the units are reserved for families eligible for Section 8 assistance. The project has been well maintained and has been 100% occupied since its construction.

The financing for Lexington Village's $13.3 million first phase was unusually complex, involving 24 separate streams of funding. The city provided $2 million in urban renewal bonds for site preparation, $2.6 million in Urban Development Action Grant loans, and $1 million in CDBG low interest loans. Additional low interest loans were made to the project by the Cleveland Foundation ($800,000), the Gund Foundation ($330,000), and the Local Initiatives Support Corporation ($250,000). Cleveland's corporate community loaned approximately $250,000, and Ameritrust Bank wrote a $2 million mortgage at a fixed rate of 10%. The city's land bank contributed parts of the site at no cost.

Another Hough CDC, Hough Area Partners for Progress (HAPP), was established in 1981 under the direction of Hough Councilperson Fannie Lewis ("Hough city councilwoman Fannie Lewis," 1996).

HAPP has facilitated two important housing development projects: Beacon Place and the Hough New House Program.

Beacon Place consists of an 11-acre site with 60 attached brick and vinyl-clad townhouses starting at $119,000 and 32 single-family detached homes with up to 2,200 square feet of space ranging in price up to $200,000. Planned in the neo-traditional style, it aims to recapture small-town America with models featuring back alleys, bay windows, and white-railed front porches. Each sale includes a 15-year property tax abatement and a 5.7%, 30-year, fixed-rate mortgage. The development is built next to a new 120,000-square-foot shopping center named Church Square, where President Bill Clinton and Vice President Al Gore announced the Empowerment Zone Program in 1994.

The Hough New House Program is making the most visible change in the neighborhood since the riots. It consists of single-family "suburban style" homes, built on large lots for middle-class owners who pre-select and build their own homes. As of June, 1996, 97 new homes had been built (or permits applied for). Prices for the new homes ranged from $65,000 to $645,000, and annual family incomes of the buyers ranged from $40,000 to $240,000.[1] Each house receives low-cost or no-cost land from the city's land bank, a forgiven second mortgage, a 15-year property tax abatement, and a below-market mortgage. As a result of these inducements, the down payment on a $150,000 home could be as low as $5,000, with monthly payments on a 30-year, fixed-rate mortgage as low as $725.

The prospects for success of all three major housing developments in Hough have been helped by the enviable location of the neighborhood. It lies midway between downtown and University Circle, the two largest job locations in Cuyahoga County. It is also only a few blocks from the Cleveland Clinic, the city's largest private employer.

## CENTRAL

The Central neighborhood is just east of Cleveland's downtown. It is bounded by East 22nd and East 55th Streets and Carnegie and South Woodland Avenues. Central is one of Cleveland's oldest neighborhoods, serving as home for successive waves of European immigrants in the 19th century and, in the years since World War I, as Cleveland's oldest, most distressed racial ghetto.

Prior to 1880, Cleveland's small black population enjoyed social and economic conditions that were unusually good for northern cities of that time (Kusmer, 1978). There was no trend toward ghettoization; no ward in the city was more than 5% black, and no segregated neighborhoods existed. Racial prejudice was not absent from Cleveland, but segregation in public accommodations and housing, common in other cities, was virtually nonexistent. Cleveland's public schools were integrated, and a substantial number of blacks enjoyed the economic opportunities of a growing economy, working as professionals, skilled tradesmen, and artisans.

By 1910, other ethnic groups in Cleveland began the process of residential assimilation and dispersion while blacks became ever more concentrated. Real estate dealers refused to show property outside the ghetto to blacks, the use of restrictive covenants increased, and many white property owners refused to sell to blacks. As blacks became more numerous, racial discrimination and segregation deepened. On the eve of the "Great Migration" of 1916-1920, racial discrimination in the sale and rental of property had become as widespread in Cleveland as in most American cities.

With limited residential opportunities and a rapidly growing population, blacks in Cleveland "piled up" in Central. By 1920, Central's population climbed over 78,000, making it the most heavily populated neighborhood in Cleveland, the home to nearly 10% of the city's population. The Central neighborhood, with its cheap lodging houses, deteriorating homes, and vice conditions, housed a majority of the black population under conditions that were decidedly inferior to those of most of the city's residential neighborhoods.

During the Great Depression, extremely high rates of unemployment and poverty, housing deterioration, and overcrowding further depressed Central. The state and federal governments responded in the New Deal with programs resulting in the nation's first public housing projects. These included Cedar Apartments (1936), Outhwaite Estates (1936 and 1939), and Carver Park (1943). Post–World War II public housing projects in Central included Cedar Extension (1954) and King-Kennedy (1970). By 1980, in an area less than a mile in radius, there were almost 6,000 units of subsidized housing, most of it owned and operated by the Cuyahoga Metropolitan Housing Authority (CMHA). It was the largest and most dense concentration of public housing in the state of Ohio. By every measure of low income, unemployment, physical deterioration, crime, and social dislocation, the Central area was Cleveland's most distressed residential neighborhood.

Central's 1990 statistics outline the extent of the social and economic disaster that plague the people of the neighborhood (Table 6.2). Median family income was $6,771—only 19% of the county median; 82% of all households were headed by single female parents (compared to 29% citywide); 56% of all adults had not finished high school; nonemployment rates (unemployed, not looking for work, and in jail) were over 50%; violent crimes (homicide, rape, and assault) were four times Cleveland's average; and only 4.3% of all homes were owner-occupied (compared with 44% citywide).

During the early 1990s, the CMHA received about $100 million above its normal budget from the federal government. The grants were facilitated by Congressman Louis Stokes, who, until 1994 and the Republican reorganization of the House, was the Chairman of the House Appropriations Sub-Committee on Public Housing. Most of the money was put to use by the CMHA in Central, where approximately 2,400 units are in some stage of comprehensive rehabilitation, costing between $70 and $100,000 per unit. Through conversion of smaller to larger units, demolition that is planned for 550 units, and conversion of about 200 units to social service purposes, nearly 900 dwelling units are planned for elimination, leaving about 1,500 of the original 2,400.

As in the Hough neighborhood, partnerships of local CDCs, philanthropic intermediaries, and private developers are building new housing in Central. Between 1990 and 1996, about 125 units of single-family detached housing were built in Central. A project that has received national recognition is the Central Commons development, an 81-unit development that emphasizes detached single-family homes set on a modified suburban style street pattern. This "urban village" is being constructed in the New Urbanism style based on a master plan designed by Andres Duany and Elizabeth Plater-Zyberk, best known for their design of upscale and neo-traditional developments.

The Central Commons project is being spearheaded by a local CDC known as the Bell, Burton, Carr Development Corp., in cooperation with New Village Corporation, the real estate development arm of Neighborhood Progress Inc., one of Cleveland's citywide intermediaries. A key source of funding for the project is a commitment of $2 million

**Table 6.2**    Central: Neighborhood Indicators by Census Tract, 1950-1990

|  | 1950 | 1960 | 1970 | 1980 | 1990 |
|---|---|---|---|---|---|
| Total population | 69,665 | 52,675 | 27,280 | 19,363 | 13,788 |
| Total housing units | 16,982 | 16,573 | 11,220 | 9,507 | 7,845 |
| Number of owner-occupied units | 1,581 | 1,413 | 613 | 530 | 338 |
| % owner-occupied units | 9.31 | 8.53 | 5.46 | 5.57 | 4.31 |
| Median housing value ($) | —[a] | 9,271 | 8,619 | 14,158 | 20,341 |
| Number of one-dwelling units | 2,119 | 3,939 | 1,361 | 1,301 | 1,138 |
| % one-dwelling units | 12.48 | 23.77 | 12.13 | 13.68 | 14.51 |
| Number of residents 5 years + | 54,765 | 19,252 | 12,019 | 10,072 | 6,147 |
| % residents 5 years + | 78.61 | 36.55 | 44.06 | 52.02 | 44.58 |
| Median contract rent ($) | 29 | 55 | 65 | 70 | 101 |
| Median annual family income ($) | 1,851 | 3,215 | 4,008 | 5,632 | 6,771 |
| Median contract rent as % of median annual family income | 18.80 | 20.53 | 19.46 | 14.91 | 17.90 |
| Median family income as % of county median | 58.71 | 46.31 | 42.49 | 25.52 | 18.94 |
| % families below poverty level | —[a] | —[a] | 40.33 | 60.78 | 73.63 |
| % households receiving public assistance income | —[a] | —[a] | 32.45[b] | 52.56 | 61.60 |
| African American population | 58,752 | 46,460 | 23,801 | 17,722 | 13,147 |
| % African American | 84.34 | 88.20 | 87.25 | 91.53 | 95.35 |
| Number of families | 16,150 | 10,815 | 5,655 | 4,273 | 2,926 |
| Number of female-headed families | —[a] | —[a] | 2,549 | 2,964 | 2,414 |
| % female-headed families | —[a] | —[a] | 45.08 | 69.37 | 82.50 |
| % high school graduate or higher | 13.61 | 15.90 | 13.38 | 27.94 | 43.91 |
| % civilian labor force unemployed | 16.55 | 17.21 | 21.47 | 37.92 | 43.39 |

SOURCE: U.S. Census data.
Note: Data are presented for census tracts 1079 (G-9), 1087 (H-7), 1088 (H-8), 1089 (H-9), 1093 (I-3), 1096 (I-5 and I-6), 1097 (I-7), 1098 (I-8), 1099 (I-9), 1103 (J-3), 1129 (L-9), 1137 (M-7), 1138 (M-8), and 1142 (N-2).
a. Data not reported by the census for this category in this period.
b. For 1970, the percentage of households receiving public assistance is reported as the percentage of families receiving public assistance (not households).

in funding from the Department of Housing and Urban Development (HUD), including $600,000 in federal Nehemiah initiative grants. The new homes are built on free, vacant land with a 15-year property tax abatement. Qualified buyers are eligible to make a 10% down payment, receive a $10,000 to $15,000 forgiven second mortgage, and obtain a general mortgage at fixed below-market interest rates for 30 years. Depending on eligibility, some families may be able to buy a $115,000 single-family home in Central Commons with monthly payments as low as $550.

In general, Central Commons exemplifies the Clinton administration's emerging "Homeownership Zone" federal urban policy initiative. This new federal program prescribes larger-scale housing developments designated for home ownership in the semi-suburban New Urbanism design style, with particular attention to projects that complement existing public housing estates (U.S. Department of Housing and Urban Development, 1995). HUD expects to dedicate $135 million nationwide in 1997 to support federal Homeownership Zones.

Developments in Central underscore a fundamental dilemma in current housing policy whereby the CMHA struggles to deal with two conflicting imperatives: (a) address the acute housing needs of the persistently poor, traditionally located in Central, and (b) follow the national consensus in favor of seeking an income mix of tenants in scaled-down developments, leaving fewer units for the poor. Low-income housing advocates are relieved at the prospect of lower concentrations of public housing but deeply concerned over the prospects of providing decent shelter for the growing number of poor families in the Cleveland area.

## EMPOWERMENT ZONES

As Central, Hough, and other poverty neighborhoods in Cleveland struggled with their problems, the federal government announced in December, 1994, that the poorest areas in the city of Cleveland would be designated a "Supplemental Empowerment Zone" (SEZ), a designation that did not exist when the Empowerment Zone program was first announced (Figure 6.1). Included in the SEZ designation were the Hough, Glenville, and Fairfax neighborhoods, with a 1990 population of about 50,000 having the following characteristics: 97% African American, a poverty rate of 46%, and a 25% unemployment rate. The Central neighborhood was not included because the program's planners believed the new housing construction and the extensive improvements under way at the CMHA would revitalize the neighborhood. The 5-square-mile SEZ area will receive a grant of $90 million over a 10-year period, plus a HUD-funded loan pool of $87 million. The program's planners estimate that another $432 million in private funds (mostly mortgage commitments) will be "leveraged" by the federal funds.

Ultimately, the city hopes the federal grants will generate up to $1 billion over the next decade. The SEZ is primarily a "place" program, attempting to restore disinvested city neighborhoods by inducing the return of population, investment, and jobs.[2]

How likely is it that the SEZ will restore the three targeted neighborhoods to social and economic viability? The $177 million Cleveland will receive over the next 10 years ($90 million in SEZ funds plus $87 million in the HUD-backed loan pool) is not much more than many existing federal programs. In 1996, Cleveland's CDBG amounted to about $33 million a year plus another $7 million for the HOME housing program; the U.S. Department of Labor provides Cleveland with an additional $10 million a year for training under the Job Training Partnership Act; and Cleveland's public housing authority has an annual budget of about $135 million including operating funds ($40 million), redevelopment money ($40 million), and HOPE VI funds ($20 million) for the 11,500 units of conventional public housing in the CMHA's 38 developments. The agency also has annual subsidies of $35 million for the 8,000 units of private housing it manages under the Section 8 program. These funds are not exclusively targeted to Cleveland's three SEZ neighborhoods, but SEZ residents make disproportionate use of them.

Although the federal government provides Cleveland with "place-bound" subsidies for neighborhood revitalization, it also provides subsidies to the region outside the city. For example, homeowners in the Cleveland region will be allowed to deduct from their federal income tax bills an estimated $500 million each year in the form of mortgage interest payments. The Cleveland region also gets about $300 million a year in federal transportation funds. Most of this money will flow into the regional highway network, facilitating continued regional suburbanization and weakening Cleveland's neighborhoods.

What about the new jobs, since jobs are the bottom line promise of the SEZ? Recent research has estimated that only 11,233 new low-skill job openings will be developed in the Cleveland region over the next 5 years (Leete & Bania, 1996). The number of projected new low-skill job openings will be far fewer than the number of job seekers. Furthermore, most of the openings are likely to be out of reach of SEZ residents. Seventy-five percent of these new jobs will be located in the suburbs and will demand long commutes by SEZ residents, 45% of whom have no car and depend on public transit for their total mobility within the

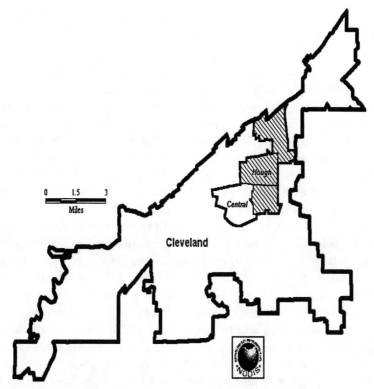

**Figure 6.1.** Cleveland's Hough and Central Neighborhoods and Supplementary EZ
(SEZ)
SOURCE: Prepared by Northern Ohio Data & Information Service, a member of the Ohio GIS-
Network, The Urban Center, Levin College of Urban Affairs, Cleveland State University.

metropolitan area. Within a 45-minute bus commute, SEZ residents will
be able to reach only 21% of the projected low-skill job openings in the
suburbs. So far, discussions with Cleveland's regional transit authority
have not resolved this issue.

Another major problem is sharply falling resources in the SEZ neighborhoods. Because of cuts in food stamps and general assistance and
welfare payments, the three SEZ neighborhoods are losing a total of
about $25.9 million a year. From June, 1996, to June, 1997, a total of
16,639 Clevelanders were cut from the food stamp program (Cleveland
Council of Economic Opportunities, 1997).

As a result of these and other realities, it seems likely that Cleveland's
SEZ will play itself out in the following scenario. The SEZ designation

will be received with much fanfare, with politicians emphasizing the probability that the three SEZ neighborhoods will now be restored. New or expanded social service agencies, like the planned three Family Resource Centers and a new Alcohol Treatment Center, will open in the SEZ neighborhoods. Using low-interest loans or grants, some existing businesses in the SEZ will expand, and a few new businesses will relocate into the SEZ. (As of November, 1996, $16.9 million had been loaned or granted to 12 EZ projects; "Federal Funds," 1996). Some residents will be trained for new jobs, but most who get the new jobs will move out of the SEZ. A substantial amount of new housing will be developed in the SEZ, but the rate of new construction will slow sharply as city-owned land in the land bank is sold off. In spite of these positive activities, racial segregation, poverty, and crime rates will remain relatively high in the SEZ, and school performance for SEZ children will remain relatively low. In 10 years, local newspapers will be running stories on the "failures" of the SEZ. It actually may not take that long; the *Plain Dealer* headlined a story "Federal Funds 'Empowering' Few Residents in Grant Zone" (Vickers, 1996).

But the newspapers will be wrong. The SEZ will not have failed; it simply will have generated the kinds of mixed results that most newspapers find difficult to address. The SEZ inducements will hold some existing firms and their jobs in the city and attract a few new firms. The SEZ also is likely to build much affordable housing, but this will have less to do with the SEZ than with the infrastructure developed in Cleveland over the last 15 years: competent neighborhood CDCs, low-income housing tax credits, strong city-bank agreements for neighborhood lending brought on by Community Reinvestment Act deals, and a large amount of city-owned vacant land. The SEZ also is likely to hold some of the city's income and property tax base in place, train some people for jobs, and provide some social services. All of this can be justified as improving conditions in SEZ neighborhoods and providing the people living there with an improved standard of living.

The SEZ is *unlikely* to deal with the fundamental problems of Cleveland's neighborhoods. It is unlikely to reduce industrial decline or get many people back to work; it is unlikely to significantly reduce concentrated poverty; it is unlikely to reduce racial segregation; and it is unlikely to improve schools or other important local institutions. To accomplish these fundamental changes, Cleveland will need a broader approach, one that tries to build on the SEZ and other *place* programs but also aims at *people-based* programs that would help remove the complex barriers that isolate residents of the SEZ and other areas of the inner city from regional economic opportunities.

A broader and probably more effective strategy would include both *place* and *people* elements. It would aim to provide jobs and income for those in Cleveland's SEZ neighborhoods who are isolated and deprived but prefer to remain where they are. It also would encourage Cleveland's minority poor to voluntarily select suburban or regional locations for jobs, housing, and schools. This more comprehensive strategy would have six elements. It would do the following:

1. Maintain and expand (to the extent possible) central city place programs such as CDBG and SEZ.
2. Provide job training to prepare inner-city workers and potential workers for jobs in suburban back-office operations, retail activities, or light industries.
3. Create job information systems, *not* through city or county Private Industry Councils (PICs) but through regional PICs, to match city workers and suburban employers.
4. Restructure public transportation systems to facilitate outward-bound journeys to work. This would mean modifying bus routes, schedules, and fares, as well as providing various forms of para-transit and/or subsidizing private automobile costs. It would mean many fewer new starts in capital-intensive rail systems designed to move suburban residents to the CBD.
5. Provide adequate day care at all stages of the process for parents of young children. (In a tight, or properly, subsidized labor market, suburban employers might be willing to share these costs.)
6. Provide adequate numbers of low-income and affordable housing units across the entire Cleveland region so that city workers will be able to translate their successes in the workplace into better residential choices and environments.

This place-and-people strategy provides needed entry-level workers to suburban employers. It provides city residents who wish to stay in their familiar neighborhoods with access to some local jobs and to the jobs being made available in the region. It maintains important aspects of African American political power and patronage in Cleveland and similar cities.

It also provides access to jobs, income, and affordable housing for those who might prefer to leave the city and relocate to better neighborhoods in the suburbs. In doing so, it weakens concentrations of poverty and may weaken racial segregation as well. The overriding idea is to improve the ghetto while also connecting its people to opportunity and social mobility wherever they exist in the metropolitan region.

## NOTES

1. Characteristics of the Hough New House Program are the following:
   - All buyers get 15-year property tax abatements.
   - All buyers get assistance with the down payment; a few have gotten a "forgivable" $20,000 second mortgage loan, and most have gotten a $10,000 second mortgage. Some banks have reduced their down payment requirements to 3% (from 10%).
   - Land costs only $100 a parcel. These parcels come from the city's land bank, which owns about 16% of all properties in Hough.
   - Average size lots for new construction are 100 × 120 feet but range up to 1/2 acre.
   - Average 1995-1996 mortgage interest rates were 6.5%, or 2 points under prime. Local banks are vigorously competing for the business because of pressures to lend in Cleveland's neighborhoods brought on by Community Reinvestment Act agreements.
   - In some cases, the city pays for landscaping and other special excavations to remove old foundations and other public improvements.
   - All new owners are black, and 23 of the first 97 owners in the program came from suburban locations.
2. The city plans to use SEZ funds for the following programs:
   - Loans of up to 90% for development that saves or creates jobs for people who live in the SEZ.
   - Loans to acquire and develop large parcels of land for new industrial, research, and other institutions.
   - Small business loans at below-market rates.
   - Short-term working capital loans to firms too new to obtain bank financing.
   - Equity investments to pay for predevelopment and construction costs of model homes to be built in the SEZ.
   - Mortgage loans as much as 1.5 percentage points below the market rate with no points, or fees, paid based on the size of the loan.

SEZ funds also will support three Family Resource Centers, one in each of the SEZ neighborhoods. These will each be governed by a board of neighborhood residents and will provide directions and access to various social services for poor families.

# CHAPTER 7

# DETROIT

## Staying the Course—Detroit's Struggle to Revitalize the Inner City

Mittie O. Chandler

Detroit holds a dubious distinction as a participant in federally funded redevelopment projects. Parts of Detroit's central city have been designated for every major urban redevelopment program since 1949, when urban renewal launched an assault against blight and slums. These areas nevertheless continue to show signs of decline. None of the programs has arrested the trends observed in Detroit's inner city. Many questions persist regarding whether the programs achieved their goals at all. Other legitimate questions concern whether the goals and objectives of the programs were realistic and whether different measures of success should be considered in assessing outcomes. An examination of the Detroit experience should suggest some directions that policymakers or practitioners should take in the future to improve outcomes and evaluations.

Similarities have been noted between the emergent Empowerment Zone (EZ) initiative and the bygone Model Cities program. Conceptually, both entail holistic approaches to urban redevelopment encompassing social, economic, and physical concerns. Other commonalities include a competitive application process, geographic targeting, collaboration between local and federal agencies, an emphasis on community participation, and local discretion in programming (Thomas, 1997). A review of revitalization attempts made in the city may help to project the potential for the Detroit EZ initiative begun in 1997.

## DETROIT'S REVITALIZATION EXPERIENCE: COMMON THEMES

Common themes run through all revitalization efforts in Detroit. Race and ethnic relations pervaded decision making in each of the redevelopment efforts (Conot, 1974; Darden, Hill, Thomas, & Thomas, 1987; Fine, 1989; Rich, 1989; Shaw, 1996; Thomas, 1997). Second, the competing demands of neighborhood development versus downtown development or community versus business are persistent. With urban renewal and Model Cities, the proximity of neighborhood developments to the central business district (CBD) essentially meant an extension of downtown development. Third, inadequate resources plagued each of the projects. The federal dollars provided by each initiative were never enough to achieve the desired outcomes. Conflict, another theme, is inherent to political processes that involve various levels of government, several agencies, citizen participation, and allocational decisions.

The Detroit revitalization experience from the 1940s to the present has been characterized by the city's leadership prominence in urban development policy making, private sector involvement, and citizen participation. To varying degrees, these characteristics are dimensions of the urban redevelopment, urban renewal, War on Poverty, Model Cities, Community Development Block Grant, and Empowerment Zone programs discussed in this chapter.

## URBAN REDEVELOPMENT

Detroit holds a unique position in the annals of central city revitalization efforts. The city is considered one of the nation's pioneers in urban redevelopment, which came to be known as urban renewal. Detroit launched the 129-acre Gratiot Redevelopment Project as part of the Detroit Plan in 1946 before the enactment of Title I of the Housing Act of 1949 provided federal funds for urban renewal. The plan was produced by a committee of city officials appointed by Mayor Edward Jeffries. The private sector, with a stake in the fate of downtown, played a role in the strategy developed for the elimination of slums and the construction of public housing. Under the Detroit Plan, the city would declare a district a redevelopment area, condemn the land, acquire parcels from current owners, clear the sites of all structures, and sell the cleared land to developers at one-fifth to one-fourth of the acquisition costs.

## URBAN RENEWAL

Title I of the Federal Housing Act of 1949 resembled the Detroit Plan's approach of condemning slum areas for resale to private developers at reduced costs. The urban renewal statute permitted the acquisition and clearance of sites and paid cities two-thirds of the difference in the costs incurred in purchasing and clearing sites and the price paid by private developers for the land.

Besides the urban renewal legislation, the Housing Act of 1950, the Housing Act of 1954, and the Housing Act of 1959 played roles in Detroit's renewal program by providing dollars for conservation and rehabilitation, relocation payments, and urban renewal planning, respectfully. Relocation policies were refined under the Housing Act of 1950.[1] With the support provided in the 1959 act, six urban renewal projects, including the original Gratiot site, were under way in Detroit by the mid-1960s. All six were within the ring formed by the semicircular Grand Boulevard.

Displacement of residents, especially African Americans, was a problem, particularly in the early stages of urban renewal. Urban renewal exacerbated the problem of low-income housing because the housing supply was reduced through demolition and not replaced with affordable housing. Between 1946 and 1958, Detroit condemned 129 acres of land at the Gratiot redevelopment site and relocated 1,950 families. Mayor Albert Cobo, elected in 1949, pushed for the speedy clearance and private redevelopment of the Gratiot land, but he cancelled the construction of planned public housing sites throughout the city (Darden et al., 1987). Early plans for the Gratiot site had included 3,600 unit of public housing, but the number had dropped to 900 by 1955. Ultimately, no public housing was built in the Gratiot/Lafayette area. By 1958, a 22-story apartment building with rents beyond the means of the former residents was built, while other phases of the project were not yet under way.

A primary purpose of both the Detroit Plan and urban renewal—to increase the city's tax base—was attained. Urban renewal, however, never managed to overcome some of the problems that occurred in other cities as well: administrative problems that led to delays, duplication of effort, cost overruns, and long periods of vacancy. Ultimately, urban renewal gave way to other policy directions as events continued to reduce its popularity: protests against the projects, new directions in federal policy embodied in the War on Poverty and Model Cities, and the civil disturbance of 1967 (Darden et al., 1987).

## THE WAR ON POVERTY

When the federal government launched the landmark War on Poverty through the Office of Economic Opportunity (OEO), Detroit was once again in a propitious position. Prior experience with other federally funded projects designed to aid poor people (Committee on Community Action for Detroit Youth and Special Youth Employment Program) and concern about hard-core unemployment in the city led Mayor Jerome P. Cavanaugh to convene a committee to attack the problem of poverty in April, 1964. At the behest of the committee, another mayoral committee, the Mayor's Committee for Community Renewal, prepared a proposal that called for coordinated and comprehensive services concentrated in poor neighborhoods, with involvement of those to be served in the planning. In June, 1964, the city's community action agency for the antipoverty program was formed and named TAP (later TAAP, Total Action Against Poverty) (Conot, 1974; Fine, 1989).

With this proposal, the city was poised when Congress passed the Economic Opportunity Act in August, 1964, prior to the presidential election that year. Detroit gained the reputation of being the best-organized city in the nation (Conot, 1974). The city received $2.8 million of a projected annual budget of $500 million for the War on Poverty (Conot, 1974). Detroit's grant was the second largest in the nation and the largest per capita (Fine, 1989).[2]

Residents of the poverty program's four target areas were represented on the 39-member Policy Advisory Committee (PAC) charged to oversee the program (see Figure 7.1). Sixteen community representatives elected from the four target areas constituted a minority on the PAC. Area advisory committees (AACS), with membership counts ranging from 36 to 52, were supposed to ensure maximum participation of the residents. In reality, the mayor exercised control by the power to appoint the majority of committee members as well as the director and deputy director of TAP. Poor people never did have control over the poverty program, nor was there any pretense that they would.[3] Cavanaugh insisted on city control of the Detroit program, rationalizing that the city charter permitted only city government to expend federal and city government funds and that no other agency was prepared to provide the 10% local share required by the OEO grants (Fine, 1989). Over time, residents lost their clout as Congress responded to appeals from local elected officials to shift program control in their direction (Beck, 1987).

Assessments of the programs offered through the poverty centers varied greatly. TAP administered a number of programs related to training and employment, and medical and dental health, as well as the Neighborhood Legal Service Center and the Small Business Development Center. The Community Development Centers (eventually known as Community Action Centers) apparently were centers of activity in the TAP target areas and generally delivered worthwhile services to the clientele. The provision of health care services was considered deficient because funding, equipment, and facilities were inadequate.

The scope of the antipoverty program's influence was limited. All poor families in the city were eligible for services, but only 44% of families in poverty lived within the target areas. About 200,000 Detroiters eligible for the services were beyond walking distance of a center. The Detroit Skills Center, a partner in the job training and placement effort, was heralded by human resources experts as among the finest operations of its type in the world. The job training and placement program reportedly had the highest rate of success in the nation among similar agencies placing the unemployed. Criticism of the job training and placement functions, however, charged that those most in need of the services were not reached and that the primary concern was with those easiest to place. According to another allegation, officials were more interested in numbers than training quality, and some training taught obsolete skills (Fine, 1989). Before TAP was discontinued, the city had the opportunity to pursue another vision partially attributed to Mayor Cavanaugh.

## MODEL CITIES

Detroit was influential in the passage of the Demonstration Cities and Metropolitan Development Act that launched the Model Cities program. Mayor Cavanaugh was the only elected official chosen to serve on President Johnson's Task Force on Urban and Metropolitan Development, set up to suggest programs and legislation for the Great Society. During his tenure on the task force, Cavanaugh recommended the designation of one demonstration city where massive allocations of federal dollars would be infused and disparate programs coordinated (Conot, 1974). The demonstration city concept was adopted by Congress, albeit in modified form. When President Johnson announced the proposal in January, 1966, it earmarked $2.4 billion for 60 or 70 cities over a 6-year period rather than $982 million for one city.[4] Ultimately, the demonstration city concept became the Model Cities program signed by President Johnson in November, 1966. In the first round of yearly funding, $400 million was distributed to more than 75 cities.

The Model City envisioned by Detroit was scaled back significantly in the light of the reduced funding awarded. The original boundaries covering the entire city were reduced to a 9-square-mile section. The Model City contained one entire antipoverty program target area and part of a second one. The area held one-fifteenth of the city spatially and one-ninth of its population (134,000), and it contained the greatest concentration of blight. All the urban renewal projects were encompassed by the Model City, making supplemental federal funds available. Poor people outside either program area got little direct assistance.

Some areas with serious social problems, such as 12th Street, where the 1967 riots occurred, were excluded.

Questions regarding the extent of benefit to eligible persons in the Model Cities area remain unanswered (Wood, 1990). It is assumed that at least half of the target area residents in Detroit participated in at least one of the numerous antipoverty programs (Fine, 1989). The report of the Kerner Commission after the 1967 riot and a statement attributed to Mayor Cavanaugh himself infer that the program had a limited reach. The Kerner Commission reported that poor people felt untouched by the federal programs. In December, 1969, Cavanaugh stated that the War on Poverty had scarcely been able to touch the lives of many of the poor (Fine, 1989).

Analysts disagree about the impact of Model Cities in Detroit, although by most accounts it was a failure. Difficulties associated with the program are irrefutable: inadequate funding, poor coordination, debilitating political conflict, inconsistent performance, and weak federal leadership (Thomas, 1997). The disagreements are based, in part, on the inability to accurately attribute responsibility for the changes observed, such as decreased unemployment or lower poverty rates. The appropriateness of evaluation criteria is another point of debate.

Prior to the 1967 civil disturbance, Detroit received praise for its implementation of Model Cities as one of the best in the country. Moderate success related to the program included realization of urban renewal programs initiated more than 15 years earlier—the West Side Industrial Project and more high-rise buildings and townhouses in Gratiot-Lafayette Park. Additional construction of downtown hotel and office structures occurred after a 25-year hiatus (Stone, 1993).

For those able to gain access, it is likely that Model Cities provided advancements in the areas of job training, employment, education, health, and family income (Fine, 1989). Using different variables, June Thomas (1997) suggests that Model Cities had a positive impact in several areas but did not lead to cumulative improvement of inner-city communities. Between 1960 and 1990, the Model Cities area lost 63% of its population and 45% of its housing units. Emergence of black political leadership is another positive outcome associated with Model Cities (Fine, 1989; Thomas, 1997; Wood, 1990). The range of antipoverty program activities, even though governing bodies sometimes overlapped and were in conflict, provided entree for some into the political sphere. On the other hand, the poverty programs are charged with coopting the leadership of civil rights organizations because these persons held positions on their boards or staffs (Fine, 1989). Model Cities ended in 1974 with the enactment of the Community Development Block Grant Program.

## COMMUNITY DEVELOPMENT
## BLOCK GRANT PROGRAM

Legislation for the Community Development Block Grant (CDBG) Program consolidated seven existing categorical programs, including Model Cities and urban renewal. Diffuse CDBG objectives were inconsistent from the start regarding the priority of program benefits to poor people (Frieden & Kaplan, 1990). Conflicting and competing objectives called for principal benefits to flow to low-and moderate-income persons *and* for undertaking activities to attract persons of higher income. Moreover, in 1977, economic development was officially added to the list of eligible activities, although some communities had previously used CDBG funds for projects with an economic development focus (Rich, 1993).

Local communities had substantial discretion to implement CDBG program objectives and to direct the use of resources within broad guidelines that varied in specificity and oversight from one presidential administration to another. CDBG is the latest on a continuum of federal programs for neighborhood activities that relate directly or indirectly to its predecessor programs. Some communities have used CDBG funds to complete urban renewal activities or to provide social services. The contest between making allocational decisions in favor of developmental or redistributive choices has been the focus of many studies (Rosenfeld, Reese, Georgeau, & Walmsley, 1995).

The advent of CDBG after both Model Cities and urban renewal signaled a change in the focus of redevelopment strategies. The social dimension of the War on Poverty was supplanted by a more physically oriented approach. Targeting of low-income persons for program benefits and citizen participation requirements also were moderated. Detroit was put in the position of making difficult choices with limited resources between the interests of neighborhood conservation and economic development (Darden et al., 1987; Orr & Stoker, 1994).

The most notable example of this quandary caused by competing needs occurred in the now infamous case of Poletown (Darden et al., 1987; Jones & Bachelor, 1986; Wylie 1989). In 1980, General Motors challenged the city to locate and deliver a cleared site within 1 year for the construction of a new plant. A task force assembled by the mayor settled on a location that held 1,362 homes and apartments, 143 businesses, 16 churches, 2 schools, and 1 hospital (Palen, 1987). General Motors bought the cleared site for $8 million, but the cost of assembly exceeded $200 million from various government pots, including CDBG, the federal Economic Development Administration, the Urban Mass Transit Authority, and state agencies, as well as a 12-year tax abatement.

The decision to develop and support the Poletown project was based on the desire to retain industry and provide jobs. It will be a while before the projected long-term benefits of the project are ever manifested in the form of the projected 6,000 direct or 20,000 indirect jobs, 4.5% increase in property tax base, and tax revenue of $21 million per year after the 50% tax abatement expires. This project will not, and could not be expected to, offset the tremendous loss of jobs experienced by the city in recent decades. The city, however, could not ignore the potential for economic growth promised by General Motors.

Decisions regarding the use of CDBG for economic development rather than neighborhood development created conflict between Mayor Coleman Young and Detroit's city council. As entrenched poverty and unemployment spread throughout Detroit, many low-income neighborhoods did not benefit from federal programs intended for low- and moderate-income persons. The federal government never allocated sufficient resources to counter the decline of city neighborhoods. Moreover, while CDBG funds remained the primary source of neighborhood and housing projects, some dollars were diverted to CBD initiatives, demolition activities, and large public and industrial development projects (Darden et al., 1986; Shaw, 1996).

The competition for funds was exacerbated by the reduction of CDBG funds going to central city communities that began during the Reagan administration (Hartman, 1986). The city received upwards of $58 million (in constant 1987 dollars) in 1985, which dropped to less than $40 million by 1990. The CDBG allocation increased modestly during the last 2 years of the Bush administration and the first 2 years of the Clinton administration (Shaw, 1996). The Young administration spent a disproportionate percentage of CDBG funds on demolition, development, and administration at the expense of increased low-income housing (Goetz, 1993; Shaw, 1996).

Mayor Young did not ignore low-income housing policy. Political contests were frequent between him and the city council over who would control the expenditure of CDBG funds to support community-based organizations and the development of low-income housing (Darden et al., 1987). Community development corporations (CDCs) had successfully petitioned city council members to confront the mayor about whether the city or community groups would administer low-income housing program. With reductions in the amount of the CDBG funds, the council maintained some measure of support for the community groups by vetoing the mayor's proposals and capturing a higher proportion of those remaining dollars. Between 1988 and 1992, the council was able to greatly reduce the proportion of low-income housing dollars that were city administered, from 35% to 16% (Shaw, 1996).

## PUBLIC HOUSING

Public housing in Detroit is run by the Detroit Housing Department (DHD), a bureau of city government. This operation has been considered as troubled by the U.S. Department of Housing and Urban Development (HUD) since 1979 and has not played an ameliorative role in providing housing for low-income persons. HUD has charged the DHD with chronically and severely mismanaging public housing resources, with a vacancy rate of 42% in the mid-1980s (Shaw, 1996).

Detroit public housing has encountered barriers of local and national origin to providing sufficient housing for low-income people (Conot, 1974; Darden et al., 1987). Insufficient dollars for construction of units, coupled with objections to dispersing units throughout the city and county, limited the growth of public housing. Major demolition and modernization projects are now under way in Detroit's public housing. Removal from the list of troubled housing authorities is pending. Although it appears that the condition of public housing will improve in the future, the number of units is likely to decline.

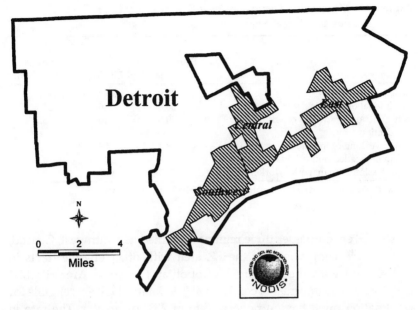

**Figure 7.1.** Detroit's Empowerment Zone
Source: Prepared by Northern Ohio Data & Information Service, a member of the Ohio GIS-Network,
The Urban Center, Levin College of Urban Affairs, Cleveland State University.

## THE EMPOWERMENT ZONE

As with urban renewal and Model Cities, Detroit is an early pacesetter
with its Empowerment Zone (EZ) initiative. The EZ application, produced by a relatively new city administration with a reorganized planning and development department, was described as extraordinary by
HUD Assistant Secretary Andrew Cuomo (Costello, 1994). The application was applauded for receiving the highest number of rating points
nationally. Commitments from the private sector were a major element
of the proposal.

Given that distress was a criterion for designation, it is not surprising
that Detroit was chosen from among 74 contenders. Economic and
social difficulties have not abated after decades of revitalization and
housing programs. The EZ boundaries, overlapping those of the Model
Cities and urban renewal areas, highlight the fact that previous redevelopment efforts did not salvage these communities (see Figure 7.1).

**Table 7.1**    Measures of Distress: Detroit Empowerment Zone and
                 Surrounding Area

| Indicator | Empowerment Zone | Eligible | Primary Metropolitan Statistical Area |
|---|---|---|---|
| % below poverty | 47.9 | 37.7 | 13.1 |
| % not in labor force | 56.5 | 47.6 | 35.7 |
| % high school graduates or with higher education | 49.0 | 57.6 | 75.7 |
| % not working or in school (ages 16-19) | 8.4 | 7.9 | 4.4 |
| % female-headed households with children | 17.9 | 21.4 | 8.6 |
| % professional/managerial workers | 45.8 | 45.1 | 59.4 |

SOURCE: Data provided by the Nelson A. Rockefeller Institute of Government.

The EZ encompasses three regions identified as Southwest, Central, and East. A comparison of the EZ area with other eligible tracts in Detroit and with the Primary Metropolitan Statistical Area (PMSA) provides a stark contrast of conditions[5] (see Table 7.1). Tracts eligible for designation must have a poverty rate of 20% or greater. The rate in Detroit's EZ is 47.9%, slightly more than 25% higher than other eligible tracts and 3.5 times as high as the poverty rate in the PMSA. In all categories, the greatest distinction exists between the zone and the PMSA. The proportion not in the labor force is 58% percent in the EZ than in the PMSA, the proportion without at least a high school educa- tion is twice as high, the percentage of 16-19-year-olds not working or in school is almost twice as high, the percentage of female-headed households with children is slightly more than double, and the propor- tion of professional workers is 23% lower.

Of the six EZ communities, Detroit ranks fourth in terms of percent- age below poverty, first in percentage not in the labor force, first in percentage with high school or higher education, second with regard to percentage of 16- to 19-year-olds not in school or employed, lowest in the percentage of female-headed households with children, and second behind New York in the percentage of professional/managerial work- ers. Detroit, then, is not dramatically better or worse off than other EZ designees.[6]

Detroit was one of six cities chosen to receive $100 million in federal Social Services Block Grant funds for a variety of purposes. The strate- gies and program activities are likely to change continually during the lengthy process of implementation. The original plan for the Detroit EZ nevertheless suggests a broad-based approach to revitalization encompassing the areas of economic development, job training, family self-sufficiency, youth, education, housing, health, transportation, other infrastructure (open space enhancement of vacant land and environmentally contaminated land), public safety, culture/recreation, and administration. Health, education, economic development, training, and other social services are to be offered by a range of providers including universities, hospitals, family service centers, and training centers.

## PRIVATE SECTOR PARTNERSHIPS

For years, the relationship between parts of the local business elite and the city administration was strained, yet in the first 2 years of the EZ initiative, private businesses reportedly invested $1.8 billion in the city. A consortium of metropolitan Detroit banks loaned $386 million in the zone, exceeding $76 million in loans promised in 1995.

Commitments from the auto manufacturers have facilitated business development and job creation in the city. Collectively, the Big Three automobile manufacturers committed $488 million and 500 jobs for the 10 years of EZ activities. General Motors (GM) sold an abandoned warehouse located on its former Cadillac Plant on Clark Street at a "favorable price" for the development of the Hispanic Manufacturing Center—a 156,000-square-foot complex of four Hispanic-owned businesses. GM also aided a business located on the grounds of the Clark Avenue facility and owned by two former Detroit Pistons basketball players that was on the verge of closing by helping it to get into another area of business. GM is also spending $250 million to expand its Detroit-Hamtramck assembly plant.

Similarly, Chrysler Corporation and Ford Motor Company have signed long-term contracts with minority suppliers located in the zone. Chrysler has completed a $1 billion project, Jefferson North, and promised to put $750 million into the Mack Avenue engine plant. In the 2 years after designation, 25 businesses moved to or expanded operations within the zone. Several of the projects were not mentioned in the EZ application.

Despite successes associated with the zone, city licensing bureaus are still perceived as obstacles to business creation and expansion, and basic service provision is criticized. Conflict between the city council and the mayor related to the EZ—attributed to the council's "us-against-them " mentality—is considered a problem by some. The empowerment zone itself is criticized as a morass of conflicting power centers and paperwork ("Challenges of 1997," 1997).

## CITIZEN PARTICIPATION

To form the EZ governing body, passage of state legislation and a city ordinance was required. The 50-member Empowerment Zone Development Corporation (EZDC) is the governing body for the Detroit EZ. Community representatives hold 60% of the seats. A quasi-public corporation, the EZDC was the product of the debate among constituencies in Detroit, with those opting for an independent body that would sustain government vacillation winning out.

The extent of ground-up and genuine citizen participation in Detroit is questioned. Despite the appointment of citizens on the EZDC, there is a strong belief that the 25-person executive committee will have the real power and will act independently of the rest of the board. Furthermore, the actual authority of the EZDC as an entity is hampered by the contract approval power maintained by the city council. The EZDC may be essentially an advisory body with a single real prerogative—initiation of change in the strategic plan, including project selection subject to the approval of the mayor and the city council (Nelson A. Rockefeller Institute of Government, 1997). The Detroit City Council maintains its power to approve all contracts with implementing agencies, so it could be perceived at the major governing force.

An early assessment of 18 Empowerment Zone/Enterprise Community (EZ/EC) projects across the country reports that the overall extent of citizen participation in the EZ/EC Initiative was significantly greater than that associated with other federal urban programs. The report points out that the struggles thus far have focused on governance structure, process, and organization, and that challenges still remain with regard to the politics of development and implementation (Nelson

A. Rockefeller Institute of Government, 1997). Guarded optimism may be an appropriate stance at this time for the city of Detroit's EZ project.

## PROGNOSIS

It appears that Detroit is off to a new start with the Empowerment Zone project. The energy created by the EZ has made some people more hopeful about prospects for the city's future. Compared to past programs, the 10-year time frame affords EZ cities the opportunity to engage in longer-term planning and implementation with a reasonable assurance of funding. EZ legislation also allows a broader approach to economic development than in the past by recognizing that other factors, such as a well-trained labor force, adequate social services, strong community facilities, equal opportunity, and freedom from crime are crucial to economic success (Thomas, 1995).

The title of Detroit's EZ application, "Jumpstarting the Motor City," has augured well for the city. In some respects, Detroit has surpassed other EZ cities. In the first 2 years, 1,750 jobs were created, compared to Baltimore's 1,600 (including jobs preserved), Atlanta's 113 created, Chicago's 87 created, New York's 407 created, and Philadelphia/Camden's 561 created. Performance in this area has been attributed to the commitments of the private sector of $2 billion, which exceeds commitments obtained in the other cities.

In some other respects, Detroit is behind other cities, particularly in enacting programs designed to improve the quality of life for zone residents. As of August, 1996, Detroit had committed $14 million of federal grants, compared to $37 million in Chicago; 17 out of 51 contracts to be negotiated for social services projects had been signed. Some delay resulted when it was necessary for the city to negotiate agreements with HUD and the Michigan Department of Human Services (now the Family Independence Agency) to use the Title XX funds. The process was protracted because appropriate rules and guidelines were not initially in place. The city has not spent much Title XX money with the agencies that will provide direct services to residents. Planning Department Director Gloria Robinson acknowledged that the average person had not begun to feel the impact of the program in the first 2 years (Stevens & Bivins, 1996). The attainment of EZ goals is more likely if obstacles such as political strife, intergovernmental conflict, and programmatic delays can be avoided.

## CAN ARCHER DELIVER?

The election of Dennis Archer in 1993, after the 20-year reign of Coleman Young, was seen by some as an opportunity to change the relationship between the city and the business community, although it was still an issue during his reelection campaign in 1997. Despite development decisions that seemed to benefit the business community, the Young administration did not enjoy a close relationship, according to some analysts (Jones & Bachelor, 1986; Orr & Stoker, 1994). Through the vehicle of the EZ, Archer may direct the city in its next step toward a renewal that is broad and sustaining.

Mayor Archer supported the establishment of the Renaissance Zone Program initiative by the state of Michigan in 1997. Among the areas chosen are six noncontiguous subzones in Detroit, totaling 1,345.7 acres, ranging in size from 67 acres to 727.8 acres outside the EZ area. The legislation provides a waiver of all state and local taxes, except the state sales tax, for residents and businesses within the zones for up to 15 years.[7] Industrial uses, and to a lesser extent commercial uses, are planned in the Detroit subzones. Five of the six subzones have residential uses and vacant residential land, but additional residential investment will not be encouraged (Sands, 1997). Nevertheless, Renaissance Zones offer another tool that Mayor Archer can use to ostensibly disperse benefits to other troubled communities.

## CAN PROGRAMS MAKE A DIFFERENCE?

The ability of revitalization efforts to have a major impact in Detroit is encumbered by the realities of regional development patterns and the state of the economy, particularly as it affects the automobile industry. The city is in tacit competition with the suburbs for development opportunities. When the so-called downtown building boom was under way in Detroit in the mid-1960s, more office space was under construction in the suburbs; 70% of industrial and 80% of commercial construction occurred outside the city. Progress within the city is over-shadowed when compared with that in the region. The industrial, commercial, and residential construction occurring at that time was primarily in the urban renewal areas, and the city was unable to dispose of additional land cleared under urban renewal in the central business district (Conot, 1974). The paradox remains evident today. In Oakland County, $1.9 billion worth of new construction and personal property was added in both 1995 and 1996. This is more than is pledged in Detroit's EZ over a decade.

Even as Detroit has succeeded in job creation since the EZ began, the number of jobs pales when compared to the suburbs. In recent decades, the city has led the region in job loss and faltered in job creation. From 1970 to 1980, manufacturing jobs declined in the Detroit (Standard Metropolitan Statistical Area) labor market, consisting of Wayne, Oak-land, and Macomb counties. Wayne County, anchored by Detroit, expe-rienced a massive loss of manufacturing jobs. Not only did Oakland and Macomb counties see increases in manufacturing jobs, but the increases in service jobs also far surpassed that in Wayne County: 10.5% increase in Wayne County, 172% increase in Oakland County, and 128% increase in Macomb County. Many of the jobs were entry level and not knowledge intensive (Wilson, 1992). Economic growth within the De-troit SMSA is occurring in areas considerably removed from the central city.

Employment is highly related to neighborhood stability. In Detroit, industrial sector employment has been pivotal to the city's well-being and that of its residents. Will the EZ help to recover from the effects of massive losses in manufacturing jobs with little replacement in their service and professional sectors? The city lost 21% of its industrial operations between 1977 and 1982, and manufacturing establishments with 20 or more employees declined by 42% from 1972 to 1982, from 821 to 477. The Detroit area continued to lose population and jobs into the 1990s. Jobs in the construction industry dropped by 8,398, or 13%, from 1988 to 1992; manufacturing jobs fell by 39,206, or 8.5%. Overall employment in the Detroit area declined by 25,000 jobs from 1988 to 1992 despite an increase in service jobs (City of Detroit, 1994). Data maintained by the U.S. Bureau of Labor Statistics report the unemploy-ment rate in the city as 16.9% in 1992; however, the rate dropped to 7.2% in 1998.

Second, Detroit's economic fate is closely tied to that of the automobile industry and decisions made by its leaders and other industrialists. Ironically, perhaps, the interdependence between the city and the auto manufacturers was as much as factor in Detroit's dramatic decline in jobs and income during the past 20 years as it is in the early stages of the EZ. Goetz (1993) questions whether local actors can shape an economy that is not dependent on the profit-based decisions of corporate leaders who are increasingly oriented toward global markets and decreasingly identified with any given locality. Detroit's experience suggests that local actors in that city have not been able to make this transition.

Early EZ activities and private sector commitments indicate that the auto industry and financial institutions are motivated to participate in the initiative. The source of motivation and dedication of the auto industry, particularly, is questioned. Orr and Stoker (1994) report that the auto industry does not see Detroit as a viable site for development, and the business community is motivated not by material incentives but, to a larger extent, by a sense of corporate responsibility or guilt about the city's condition. The rationale for corporate support may not be as important as its endurance. The incipient partnerships and commitments must be maintained throughout the 10-year EZ initiative and beyond if the city is to experience a sustained revival.

If the federal government and local officials are attentive to the development of evaluation strategies and designs, reliable findings should result from the various EZ programs. Advance planning can prevent the problems associated with the evaluation procedures and outcomes assessments of the Model Cities program and make the EZ a more useful learning experience.

## THE CITY AND SUPPORT OF POOR NEIGHBORHOODS

The city's choice of EZ boundaries bolsters previous assertions that the city has chosen to target many of its resources to specific areas such as the CBD and the corridor between the CBD and Wayne State University/Medical Center also known as the Central Functions Area. Although the EZ proposes to spend money on neighborhood redevelopment and people redevelopment as well as economic development, the earliest production results are in the latter category. The EZ contains an element of human services attendant to social problems that was lost when CDBG was initiated, but questions remain as to whether the city will be successful in maintaining the balance of activities and how political decision making and parlaying will affect the outcomes.

## NOTES

1. Federal legislation passed in 1970 set minimum levels for relocation benefits.

2. See Conot (1974, p. 500) for an accounting of how the design of Model Cities (originally called Demonstration Cities) changed from directing a concentration of funds to a few cities to dispersing $400 million to more than 75 cities.

3. Representation of poor people at 34% was higher in the Detroit community action agency than elsewhere in 1966. Over time, poor people did gain majority membership in San Francisco, Newark, and Syracuse (Fine, 1989).

4. In December, 1965, Detroit submitted a proposal to Washington for a Demonstration City based on a request from yet another task force. The proposal, assembled in 3 weeks, sought $982 million over a 10-year period.

5. The population in the EZ is 101,279; it contains 18.35 square miles of land. The racial composition is 67% African American, 24% white, and 9% other racial groups; 11% of the population is Hispanic (which can include black or white persons).

6. Atlanta, Baltimore, Chicago, New York, and Philadelphia/Camden are the other EZ communities.

7. The exclusion includes personal income tax, single business tax, state education tax, real property tax, personal property tax, local income tax, and utility users tax. In addition to Detroit, five other urban areas, three rural zones, and two military bases were chosen as Renaissance Zones.

# CHAPTER 8

# EAST ST. LOUIS, ILLINOIS
## Promoting Community Development Through Empowerment Planning

Kenneth Reardon

## THE ECONOMIC COLLAPSE OF EAST ST. LOUIS

East St. Louis, once a thriving Illinois riverfront community of bustling rail yards, packinghouses, and steel mills, known as the "Pittsburgh of the West," is now one of America's poorest cities. Its economy collapsed following World War II, when refrigerated trucks and rail cars eliminated the need for regional meat-packing centers, cars and trucks eclipsed railroads as the primary mode of travel, and Midwestern manufacturing plants moved to the South and overseas seeking lower production costs. Between 1970 and 1990, these changes led to a decline in the number of East St. Louis businesses from 1,527 to 383 and a drop in individuals employed by these firms from 12,423 to 2,699. These trends caused the value of East St. Louis's real estate tax base to fall from $560 million to $142 million, forcing the city to eliminate all but essential police, fire, and public works services. By 1990, despite repeated cuts in services and increases in property taxes in the 1970s and 1980s, the municipal debt was approximately $88 million.

The combined effects of shrinking employment opportunities, deteriorating housing conditions, rising property taxes, declining municipal services, and increasing rates of violent crime prompted many residents, especially white working- and middle-class families, to abandon the community. Between 1960 and 1990, the city's total population fell from 88,000 to 43,000, the proportion of white residents dropped from 55% to less than 2%, the proportion of female-headed households rose from 21% to 62%, and the proportion of families living in poverty increased from 11% to 39%. The city's economic problems peaked in 1991, when the city failed to make a court-mandated payment to the estate of an individual injured in the municipal jail and the presiding judge attempted to satisfy the judgment by awarding the deed to the East St. Louis City Hall and 220 acres of municipally owned riverfront property to the estate.

Such problems prompted a U.S. Department of Housing and Urban Development (HUD) official to describe East St. Louis as "the most distressed small city in America" (quoted in Kozol, 1991, p. 7) and led an editor from the *St. Louis Post-Dispatch* to refer to the community as "America's Soweto" (quoted in Kozol, 1991, p. 8). This coverage, reinforced by residents' demands for reform, prompted the Illinois legislature to pass the Illinois Distressed Cities Act of 1991, which authorized the state to provide East St. Louis with up to $34 million in supplemental low-interest credit to reduce and restructure its debt. The act also established the East St. Louis Financial Advisory Authority (ESLFAA) to monitor the city's financial activities, with the power to approve all municipal budgets, hiring decisions, and city contracts. In 1991, the state of Illinois also awarded East St. Louis a coveted riverboat gaming license. East St. Louis's *Casino Queen* riverboat has been one of Illinois' most profitable gambling establishments since it began operation in the spring of 1992. The boat generates $10 to $12 million in annual gross receipts taxes for the city, and 25% of its 1,200 jobs are held by residents of the two-county Metro East region. The *Casino Queen's* ongoing success recently led its St. Louis investors to purchase an additional 55 acres of East St. Louis waterfront property, where they plan to build, by the end of 1999, a hotel, restaurant, entertainment complex, and recreational vehicle park, at a cost of $40 to $50 million.

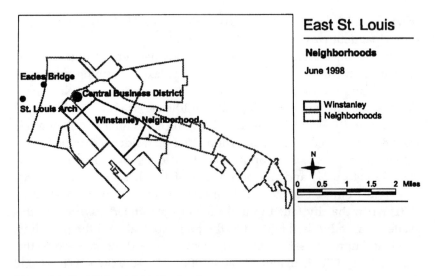

**Figure 8.1.** East St. Louis Neighborhoods
SOURCE: Map created by Deanne Koenigs, East St. Louis Action Research Project, University of Illinois at Urbana-Champaign.

The city's economic development efforts also received a boost from the federal government after Bill Clinton's election as president in 1992. The city received substantial funding increases in its Community Development Block Grant (CDBG) and other federal programs and obtained designation as an "Enterprise Community."

Despite these positive developments, the devastating impact from post–World War II deindustrialization, disinvestment, and suburbanization remains in the city's older residential neighborhoods. Established neighborhoods, especially those adjacent to East St. Louis's declining central business district, experienced a significant loss of retail stores and a skyrocketing number of abandoned buildings and vacant lots. Local government efforts to stabilize these neighborhoods only concentrated the poorest of the city's residents in high-rise public housing complexes (Figure 8.1).

One of the city's inner ring neighborhoods most severely affected by its ongoing economic slide was the 120-block area known as Winstanley/ Industry Park, which was a vibrant residential neighborhood in 1970, housing 13,640 individuals in a mix of 3,982 attractive bungalows, two-story duplexes, and small apartment buildings. By 1990, Winstanley/

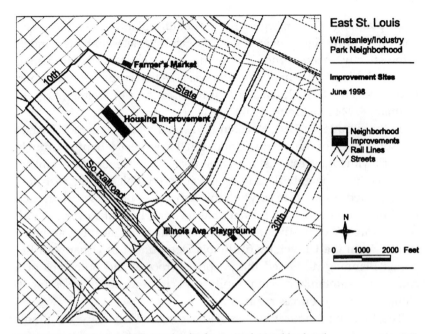

**Figure 8.2.** East St. Louis Winstanley/Industry Park Neighborhood
SOURCE: Map created by Deanne Koenigs, East St. Louis Action Research Project, University of Illinois at Urbana-Champaign.

Industry Park had lost 41% of its population and 29% of its housing, and 21% of its building lots were vacant (Figure 8.2).

## INITIATING THE EMPOWERMENT PROCESS IN WINSTANLEY/INDUSTRY PARK

The Winstanley/Industry Park neighborhood's escalating unemployment, housing abandonment, public safety, and municipal service problems prompted Rev. Gary Wilson, pastor of the Wesley Bethel United Methodist Church, to request planning assistance from the Department of Urban and Regional Planning at the University of Illinois at Urbana-Champaign in 1991. The university responded by offering to work with neighborhood residents and institutional leaders to complete a comprehensive neighborhood stabilization plan. Aware of residents' anger over the past failures of many "top-down" planning processes, the participating faculty suggested an empowerment model of neighborhood planning. This emerging form of equity planning seeks to overcome the knowledge, power, and ideological obstacles that often undermine effective neighborhood planning by drawing on the principles and methods of participatory action research that empowers neighborhood residents to seek feasible solutions to their problems (Argyris, Putman, & McLain Smith, 1987).

In the summer of 1991, Reverend Wilson invited Reverend Herman Watson, pastor of the nearby Mount Sinai Missionary Baptist Church, to work with him in establishing a Steering Committee for a Winstanley/ Industry Park Neighborhood Organization (WIPNO). They recruited 12 interested residents to serve on this temporary board, which invited neighbors to attend an informational meeting held at Reverend Wilson's church. Twenty-five individuals attended this meeting, at which faculty from the University of Illinois Department of Urban and Regional Planning discussed the purpose, process, and outcomes of a complete a preliminary community capacity and needs assessment, along with the resources required.

The Steering Committee, assisted by eight University of Illinois graduate planning students, developed a six-part research design to assess the strengths and weaknesses of the neighborhood. During the fall of 1991, local residents and university students worked together to develop a demographic profile of the neighborhood using 1960, 1970, and 1980 U.S. Census data; survey current land uses and building conditions; evaluate the state of local infrastructure; analyze property ownership and tax payment trends; interview local institutional leaders regarding their perceptions of the area; and query neighborhood residents about their goals for the neighborhood. In May, 1992, more than 100 community residents attended an information meeting organized by the Steering Committee to review the preliminary results of the neighborhood assessment. The committee's presentation highlighted both positive and negative aspects of the neighborhood.

## FORMULATING NEIGHBORHOOD GOALS FOR WINSTANLEY/INDUSTRY PARK

The joint resident/student committee strongly encouraged the neighborhood to adopt the following seven community stabilization goals.

1. Enhance public health and safety by implementing an immediate infrastructure repair and housing demolition program;
2. Stabilize the existing residential housing stock by reducing building operating costs and assisting owners in making needed repairs;
3. Improve the appearance and functionality of the neighborhood;
4. Expand local business activity and employment opportunities;
5. Reduce alcohol and drug abuse and related criminal activity through prevention and treatment programs;
6. Pursue citywide and regional policies to make corporations and government more responsive to the needs of older residential neighborhoods; and
7. Empower local residents to address their pressing needs by establishing a permanent neighborhood organization.

Four graduate planning students agreed to assist WIPNO in accomplishing the following objectives during the 1992 academic year: (a) prioritize neighborhood needs; (b) develop an integrated program of housing improvement, economic development, and community building; (c) expand resident recruitment efforts to build an active membership base for the new community development corporation; (d) create a more formal (501(c)3) organizational structure; and (e) identify available funding and technical assistance programs to support local neighborhood improvement efforts.

The Steering Committee, assisted by these graduate students and 50 undergraduate volunteers, conducted a door-to-door survey of 550 Winstanley/Industry Park households to determine their neighborhood improvement priorities. An analysis of the survey results showed strong community support for an aggressive outreach campaign to increase resident participation, a comprehensive program to improve local housing conditions, and a targeted effort to expand employment and business opportunities for longtime unemployed neighborhood residents. The committee responded to these results by recruiting 45 residents to participate in three issue task forces charged with exploring innovative solutions to these problems.

## CREATING THE ILLINOIS AVENUE PLAYGROUND

Soon after establishing its issues task force, the WIPNO Steering Committee was approached by a small group of senior citizens requesting assistance in creating a toddlers' playground for their area, which was located approximately 10 blocks south of the Winstanley/Industry Park neighborhood. Respectful of the seniors' decades of community service, impressed by their commitment to the project, and eager to demonstrate its effectiveness as a community development organization, WIPNO agreed to work on the project despite its location outside the Winstanley/Industry Park neighborhood.

In the fall of 1992, a small group of local seniors, Steering Committee members, and university students met at the proposed playground site to discuss the means (design, fund-raising, construction) required to complete the project. Consensus on a preliminary set of design guidelines was quickly achieved as steps were taken to secure site control of five abandoned building lots from the St. Clair County Trustees, who were responsible for tax delinquent properties. Efforts also were launched to secure $15,000 worth of plant materials, construction equipment, and building supplies for the project from area nurseries, lumber yards, do-it-yourself stores, and University of Illinois alumni.

As work progressed, one of the participating residents criticized the committee for not involving the neighborhood's most knowledgeable play "experts," the local children for whom the playground was being built, in the design process. Embarrassed by this oversight, the committee invited area children interested in helping to design the playground to show the group their favorite neighborhood play spaces, which included an abandoned factory and a cemetery, and to share their image of the "ideal" playground on a 40-foot piece of paper. The children were then assisted by a landscape architecture student from the university in organizing their most popular playground proposals into a site plan for the project. When the children's design proposal was presented to WIPNO and two area church groups, local commitment to the project soared.

This commitment was quickly tested when the St. Clair County Trustees said they would not transfer the titles of the five tax delinquent properties selected for the toddlers' playground to WIPNO until it (a) received nonprofit status from the state of Illinois and tax-exempt recognition from the Internal Revenue Service (IRS), (b) secured written forgiveness of all past-due taxes on these five parcels, (c) obtained a $500,000 general liability policy to protect the organization from any civil suits, and (d) designed and implemented a maintenance program for the playground's upkeep.

Undeterred by the county's decision to hold WIPNO accountable to a set of development standards not applied to other nonprofit organizations or municipal agencies, the Steering Committee worked to meet the requirements. In April, 1993, WIPNO received deeds for three of the five lots from the county, which made it possible for a modestly scaled down project to proceed.

In May, 1993, 100 volunteers, supervised by two local African American contractors, used plans developed by university planning and design students based on a site plan created by two dozen East St. Louis children to transform three trash-strewn lots into a beautiful 23,000-square-foot playground. Sounds of laughter now echo from the Illinois Avenue Playground, which features a double Dutch platform, a sandbox, a climbing structure, a tire maze, walking paths, and several planting beds containing dozens of fruit trees, hundreds of shrubs, and thousands of flowers. It has been maintained since its completion, without assistance from the city, by the senior citizens who proposed this project and the children and parents who helped build it. Not one piece of playground equipment has been vandalized or stolen.

## DEVISING THE NEIGHBORHOOD STABILIZATION PLAN

As work progressed on the Illinois Avenue Playground project, Steering Committee members, assisted by four graduate planning students, collaborated on the development of a neighborhood stabilization plan focused on jobs, housing, and community. The resulting WIPNO Community Development Strategy was designed to create a sustainable community revitalization movement involving ever-increasing numbers of neighborhood residents and local institutional leaders. The committee proposed a series of open space and residential housing improvement programs to address the neighborhood's "place-related" problems, a variety of human resources and capital development efforts to resolve the neighborhood's "labor-market" problems, and a mix of citizen participation and human capital initiatives to remedy the neighborhood's "people-centered" problems.

Aware of the widespread cynicism engendered by the failure of many past planning efforts, the committee proposed a 5-year time line that progressed from relatively simple to increasingly complex projects.

Using this developmental approach, WIPNO's leaders hoped to use the momentum generated by a series of successfully implemented, albeit modest, early projects to build a broad base of resident and official support for its efforts. The leaders believed that small-scale projects successfully completed on a voluntary or self-help basis would encourage public and private community development agencies to fund future efforts.

Eager to demonstrate the impact that a coordinated community development plan, executed on a developmental basis, could have on the quality of life in a single neighborhood, the leaders decided to concentrate their neighborhood improvement efforts within a 12-block target area. The plan identified the largely residential area surrounding Wesley Bethel United Methodist and Mount Sinai Missionary Baptist churches as WIPNO's first neighborhood demonstration area. Although this area suffered from a significant number of abandoned homes and vacant lots, it also benefited from a large number of well-maintained owner-occupied units and the interest of these two influential religious congregations. WIPNO's Steering Committee began an intense fund-raising effort immediately following the 1993 adoption of its Five-Year Neighborhood Demonstration Area: Strategic Community Stabilization Plan. Various grants enabled WIPNO to rent space for an office, purchase needed equipment and supplies, and hire a full-time planner.

## ESTABLISHING THE EAST ST. LOUIS FARMERS MARKET

Local job generation became the new planner's first priority immediately following his appointment in July of 1993. During the summer of 1993, he worked with WIPNO's Economic Development Committee to design a household spending survey that was completed by University of Illinois planning students in the fall of 1993. It revealed that food expenditures significantly exceeded housing-related costs for the majority of East St. Louis households. These results, along with encouragement from the head of the University of Illinois' landscape architecture department, led WIPNO's Economic Development Committee to explore the redevelopment potential of productive agriculture and food retailing in East St. Louis. The committee hoped to expand employment and business opportunities for residents through a "buy-local" campaign encouraging local consumers to spend more of their food budgets on locally grown and locally marketed fresh fruits and vegetables. Research by Michelle Whetten, a graduate planning student, identified suburban food purchases by East St. Louis residents as a major source of lost revenue to the local economy and highlighted a dramatic drop in East St. Louis retail food establishments, from 75 in 1972 to 13 in 1979.

The committee quickly abandoned its productive agriculture plans in favor of food retailing when an investigation of East St. Louis soil contamination completed for WIPNO by the University of Illinois Department of Nuclear Energy revealed very high levels of heavy metal contaminants at several proposed East St. Louis garden sites. Other possibilities, however, were explored. In the fall of 1993, with the assistance of several planning students, a literature review of public food markets was completed and field trips to two wholesale and six retail markets in nearby cities were conducted, producing a wealth of food market information.

This research led WIPNO's Economic Development Committee to pursue the establishment of a community-owned and -managed retail vendor food market in East St. Louis. Assisted by Ms. Whetten and myself, the committee developed a 15-page development proposal for an East St. Louis Farmers Market, which it asked 45 private corporations and public agencies to fund. Only the East St. Louis Community Development Block Grant (CDBG) Agency and Business Development Office (BDO) were willing to provide financial support for the project. In the spring of 1994, the city's CDBG Agency provided WIPNO with a $30,000 grant to purchase a market site in or near the central business district. The BDO then gave WIPNO a $30,000 "soft" loan, to be forgiven over 5 years, to enable the organization to erect a simple market "shed" at its selected site.

Shortly after securing project financing for the farmers' market, WIPNO acquired a 80,000-square-foot site on the city's main commercial corridor, within 10 blocks of the core of the city's central business district. At the committee's request, the university's planning students investigated the legal requirements of operating an outdoor food market, initiated a recruitment campaign to recruit market vendors, and designed a small business training program for inexperienced vendors. Landscape architecture students developed an overall site plan that

provided excellent automobile and pedestrian access, convenient cus-
tomer and merchant parking, and attractive planting beds. Architecture
students devised renovation plans to transform an aging carport on the
site into an attractive market shed. Local residents and campus volunteers then completed more than 43,000 hours of work to transform the
abandoned used car lot into an attractive public market site.

Two hundred WIPNO members, local residents, municipal officials,
and student volunteers cheered the opening of the East St. Louis Farmers Market on May 4, 1994. During its first three seasons of operation,
the market generated more than $395,000 in local retail sales and $76,500
in employee wages for local residents. These revenues, along with their
local multiplier benefits, helped East St. Louis recapture more than
$961,824 worth of business activities from its surrounding suburban
communities. The market's ongoing schedule of food demonstrations,
nutrition lectures, health screening fairs, and musical programs makes
it an important center for community activity. The new shoppers the
market has attracted to the State Street business corridor have encouraged several nearby retailers to improve their facades and expand their
product offerings.

## IMPROVING WINSTANLEY/INDUSTRY PARK'S HOUSING STOCK

As work progressed on the farmers' market in the fall of 1993, WIPNO's
Housing Committee explored various strategies for improving the
neighborhood's building stock. The abysmal community reinvestment
ratings of several local lending institutions, along with HUD's decisions
to place the management of the East St. Louis Public Housing Authority
and Community Development Block Grant Agency in the hands of
private receivers, led WIPNO to consider various self-help approaches
to housing improvement. In the fall of 1993, WIPNO identified dozens
of single-family homes owned by low-income senior citizens in need of
minor exterior repairs and painting. Working with students from the
University of Illinois and Wartburg College in Waverly, Iowa, lumber-
yards, paint stores, and home improvement centers were solicited for
donations of basic materials. The exteriors of more than a dozen senior
citizens' homes were scraped, primed, and painted by students during

the first year of WIPNO's volunteer Paint and Scrape Program. A similar number of homes have been improved during each of the past 5 years.

The success of this volunteer home improvement program brought WIPNO to the attention of the Metro East Lenders Group, a coalition of area banks formed to promote community reinvestment lending in East St. Louis. The group contacted WIPNO to secure its help in increasing the number of East St. Louis residents and businesses seeking mortgages, home improvement loans, or lines of credit from their institutions. WIPNO proposed increasing local bank applications through a coordinated community education, local media, and neighborhood outreach program. A scaled down version of the campaign was implemented by WIPNO in 1994 through a $25,000 grant from this banking consortium. More than a dozen families, most of whom were burdened by usurious bond-for-deed contract payments, were able to secure conventional bank financing in 1994 through this credit counseling program, which won a national award for WIPNO and the participating banks.

WIPNO also explored various public programs for home improvement. In the winter of 1993, WIPNO contacted the Illinois Treasurer's Office to inquire about the new "direct deposit" program through which state funds are placed on deposit with local lenders who, in turn, agree to contribute the difference between the market interest rates such deposits typically accrue and the slightly lower interest rate accepted by the state to a locally administered affordable housing fund. In response to WIPNO's request, the state treasurer placed $1 million on deposit with Magna Bank's East St. Louis branch. The interest payments the state deferred on this deposit over 3 years generated a $75,000 affordable housing loan fund that was used to help many East St. Louis families with home improvement loans, closing cost subsidies, and home mortgage down payment assistance.

WIPNO's growing reputation as an effective housing improvement organization prompted local officials to encourage its leaders to seek federal funding for single-family home rehabilitation under HUD's newly launched HOME Program. With the assistance of members of the University of Illinois architecture and planning faculty, WIPNO applied for $175,000 in HOME Program funds to rehabilitate the homes of seven low-income senior citizens. In 1995, WIPNO used these funds to substantially improve six of the seven homes. This led to WIPNO's designation as a Community-Based Housing Development Organization (CHDO) by the East St. Louis Community Development Block Grant Agency. That designation resulted in a $190,000 grant to rehabilitate two abandoned multifamily apartment buildings for rent to low-income individuals and families.

## BUILDING ORGANIZATIONAL
## CAPACITY FOR WIPNO

In the fall of 1992, WIPNO's Steering Committee recruited 75 neighborhood families as new members, several of whom worked with WIPNO's planner to draft a new constitution and set of bylaws for the organization. WIPNO's board, with pro bono legal assistance from Mr. Joseph Pavia of Champaign, then secured nonprofit status for the organization from the Illinois Secretary of State's Office and 501(c)3 recognition from the IRS.

The organization's new legal status enabled it to pursue charitable funding from many quarters. Between 1993 and 1995, the organization's operating budget increased from $45,000 to $300,000. These additional resources allowed the organization to hire a credit counselor and economic development/community organizing specialist. In 1996, WIPNO secured a $25,000 conventional mortgage from a local lender to purchase an abandoned two-story parsonage owned by the Mount Sinai Missionary Baptist Church to house its expanding operations and those of the East St. Louis Community Action Network (ESLCAN), a newly organized citywide coalition of neighborhood organizations that included WIPNO.

## PREDICTING THE FUTURE OF LOCAL
## COMMUNITY-BASED PLANNING EFFORTS

WIPNO's success in carrying out a series of increasingly complex and challenging housing improvement and economic development projects has inspired other East St. Louis neighborhoods to launch similar efforts. Three other neighborhoods have completed community stabilization plans with university assistance since 1993. One of these neighborhoods recently established its own community development corporation and has secured more than $250,000 in federal funds for housing improvement and neighborhood beautification. These resources have enabled this fledgling neighborhood organization to hire a full-time community planner and part-time administrative assistant. The new CDCs are currently receiving legal assistance from Land of Lincoln Legal Services, the University of Illinois' Legal Clinic, and the Neighborhood Law Offices. Meanwhile, they also have benefited from more than $50,000 in federal environmental protection expenditures channeled through the University of Illinois' East St. Louis Action Research Project.

WIPNO's effectiveness as a community-based housing improvement and economic development organization has made it an increasingly important partner for the East St. Louis CDBG Agency. This office appears eager to finance WIPNO's future housing and community development activities, as does the East St. Louis Public Housing Authority and Enterprise Communities Program, provided that the organization continues to complete its projects in a cost-effective manner. Several private developers also have approached the organization with various partnership proposals.

The biggest threats to this growing citizen-initiated and university-assisted grassroots development movement appear to be internal in nature. As WIPNO's housing and community development activities have expanded, its leaders and staff have devoted less time to community outreach and organizing activities. As a result, resident participation in WIPNO has fallen in recent years, leaving the organization with fewer volunteers than it needs to pursue its ambitious community development agenda. These membership losses have placed enormous burdens on WIPNO's small core of dedicated leaders, causing some to withdraw from the organization and discouraging others from stepping forward to assume greater responsibilities. The organization's shrinking leadership core has also caused some local officials, who chaff at WIPNO's political independence, to question how well the current group of officers and board members represents the neighborhood. The negative experiences many WIPNO board members have had in dealing with various municipal, state, and federal bureaucrats have made them highly suspicious of most urban professionals, and this has produced considerable tension between WIPNO's elected officers and its paid staff. This tension has also contributed to a high rate of turnover in the executive director's position. WIPNO employed four different executive directors in the period between its formal incorporation in 1991 and 1997. Turnover in this key position has disrupted the organization's programs and has raised questions regarding the group's stability. WIPNO's internal problems, if unaddressed, may undermine its ability to secure ongoing funding.

Aware of many of the organizational challenges confronting both newly established and maturing community development corporations, the East St. Louis CDBG Agency recently awarded the University of Illinois' East St. Louis Action Research Project (ESLARP) a $190,000, 2-year grant to establish a Neighborhood Technical Assistance Center (NTAC). Combining these funds and $100,000 in Community Outreach Partnership Center resources from HUD's Office of University Partnerships, the University of Illinois opened the NTAC to provide organizing, planning, and design assistance to individuals, neighborhood organizations, religious institutions, local businesses, and municipal agencies pursuing local community development projects. The NTAC received 68 technical assistance requests in its first 3 months of operation alone. Its success prompted the local Enterprise Communities Program to budget an additional $25,000 a year and to request $25,000 in matching funds from St. Clair County to enable the NTAC to provide technical assistance services to the four suburban communities that, along with East St. Louis, constitute the EZ's community service area.

## CONCLUSIONS

Although East St. Louis continues to face serious economic and financial problems, it has managed, with the help of a considerable amount of state and federal aid, to stabilize its local economy. This improvement in its financial condition has enabled the city to shift its attention from issues of crisis management to economic development, and its current revitalization strategy has shifted from one exclusively focused on large-scale downtown and waterfront projects to one that attempts to balance the needs of these commercial areas and those of its aging residential neighborhoods.

East St. Louis, for the first time in its history, is attempting to pursue an ambitious neighborhood revitalization program in partnership with the city's expanding network of community-based development organizations. The number of active neighborhood improvement organizations increased from 5 in 1990 to 15 in 1998 as residents have seen organized neighborhood groups secure resources to improve their areas in ways that meet their own neighborhood goals.

To maintain the effectiveness of this partnership, the city must ensure ongoing support for city/neighborhood joint ventures among local elected officials, and it must leverage private investment in these neighborhoods using the city's CDBG funding. The future viability of this partnership, however, will depend largely on the continuing strength of these neighborhood organizations, which will enable them to remain active partners. These organizations, most of them new and developed with the assistance of the ESLARP, will require ongoing support.

The city has recognized this in its use of CDBG funds to enable ESLARP to establish the Neighborhood Technical Assistance Center to provide support to community groups with full-time staff in East St. Louis in addition to those 3 hours away in Urbana-Champaign. Both the city and the university hope that the NTAC eventually will become an independent nonprofit with a board of community residents funded by the city, which would effectively institutionalize a city policy of support for neighborhood organizations—and the empowered residents who define their goals.

# LOS ANGELES

*Borders to Poverty—*
*Empowerment Zones and the*
*Spatial Politics of Development*

Ali Modarres

## INTRODUCTION

From the antebellum reform movements focusing on a moral diagnosis
of poverty to the most recent "war on poverty," policymakers and the
public perception they share have been entrenched in ideological war-
fare. The inherent nature of the "contact-and-removal" agenda of the
past century and the current space-based poverty programs both incorporate the idea that certain urban spaces (i.e., urban cores, where
minorities and immigrants are disproportionately housed) contain
mysterious agents of social ills and as such are conducive to poverty.
Equipped with the logic of social and spatial convergence, academic
literature and social policy have attempted to eliminate poverty by
providing means of increasing spatial mobility. From this perspective,
eradication of poverty would equate the removal of the poor from

disadvantaged areas. This is exactly what the anti-poverty policies of the last three decades have accomplished. These "urban solutions" typically provide access to funding for a short period of time, and in the past they emphasized redevelopment of blighted areas. Urban litera- ture has successfully documented the problems caused by this type of approach, and over the last decade the direction seems to have been diverted toward community development, especially business incen- tives and job training programs.

In Los Angeles, the Community Redevelopment Agency (CRA), established in 1948, represents the earlier tradition, and the recent Supplemental Empowerment Zones (SEZ) exemplify the latter. To the extent that the new tradition attempts to build on community assets, including human capital, it departs from the top-down policies of the past; however, under most scenarios, including the Rebuild LA effort, which emerged after the 1992 riots, lies the philosophy of "business knows best." In that regard, the efforts advanced under this new tradi- tion have materialized largely in the form of economic development and lack any coherent human capital development. This is not to say that job training programs do not occur, or that small business loans do not produce a few jobs. These initiatives, however, fail to produce large-scale changes in employment or, for that matter, poverty patterns. In this chapter, I will attempt to illustrate that the newly created SEZ and its single solution, the Community Development Bank, are merely another initiative in the chain of policies focused on building commu- nities through the guise of economic development. If they are anything like their predecessors, their chance of success is minimal.

## SUPPLEMENTAL EMPOWERMENT ZONE (SEZ) DEFINED

The recent war on poverty, which has been embodied by the Empow- erment Zone, has attempted to address the connectivity between eco- nomic and community development through the four principles of economic opportunity, sustainable economic development, commu- nity-based partnerships, and strategic visions of change. This policy generated much hope in the city of Los Angeles, where everyone thought that the 1992 urban riots, along with a series of social and natural disasters, would surely make the area the most deserving of federal dollars. With this belief in mind, the City and the County of Los Angeles (after an initial period of working separately) submitted a joint proposal to earn the EZ/EC (Empowerment Zone/Enterprise Community) designation for a number of their inner-city neighborhoods. Much to the embarrassment of the mayor, county officials, and other participants, Los Angeles lost its bid in the original application process and had to await the Economic Development Initiative monies and the compromise status of "Supplemental Empowerment Zone." The inadequacy of the Los Angeles application and its lack of private sector support ensured its failure. In other words, Los Angeles's self-importance prevented it from seeing the seriousness of this competition.

Probably one of the most troublesome steps in the initial process was the selection of tracts to be nominated for EZ status. After numerous public hearings and closed-door cartographic exercises, Los Angeles decided on 41 census tracts (less than 2.5% of all tracts in the county) that closely matched the federal perception of urban poverty. These tracts fall in some of the poorest neighborhoods in the City and County of Los Angeles (see Figure 9.1). Los Angeles County has more than 9 million inhabitants (Department of Finance, 1996), occupying more than 4,000 square miles of land, 85 incorporated cities, 130 identifiable communities, and a population of more than 1.3 million at or below the poverty line. The final SEZ housed slightly more than 81,000 individuals. The overall low-density characteristics of Los Angeles made it difficult to create an image comparable to the East Coast–style urban ghettos where the poor are concentrated. The multicultural characteristics of the urban poor created another layer of problems for Los Angeles: There are more Latinos at or below poverty levels in Los Angeles than African Americans (8.6% versus 2.3%, or 744,383 versus 203,286 persons).

According to the initial proposal, the tracts were chosen primarily for their poverty status, but issues such as development potential also were considered. Interestingly, the initial application under which these areas were selected included Title XX funds, so issues of capacity building were more directly considered; however, in the transition from EZ to SEZ and the related loss of social infrastructure money, the development program shifted to traditional trickle-down economics, focusing on job development. This is to be accomplished by lending

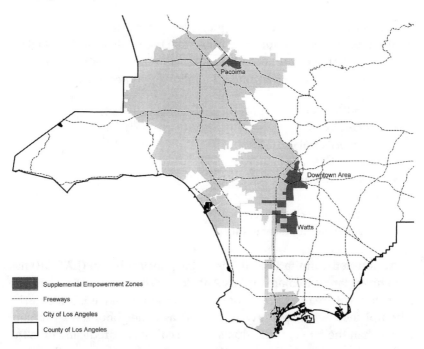

**Figure 9.1.** Los Angeles and Its Supplemental Empowerment Zones
SOURCE: The Urban Environment Initiative Data CD, Center for Spatial Analysis and Remote Sensing, California State University, Los Angeles, December 1997.

money to business owners, who will in turn hire 51% of their employees from the SEZ neighborhoods. There are no commitments to certain types of jobs, however, and no mention of how long this policy is to be implemented by individual loan recipients. This approach seems more palatable to a mayor who believes "business knows best!"

Initially, Los Angeles was less than excited about accepting the consolation prize. When the Economic Development Initiative (EDI) funds were offered 2 weeks prior to the final announcement of the EZ designations, however, the city and county agreed to be announced as an SEZ and receive $430 million. For that, Los Angeles proposed a single solution: a community bank (The Los Angeles Community Development Bank, or LACDB). The total funds available to the bank consisted of $115 million in EDI funds, $15 million of which is targeted for the few areas that fall in the county jurisdiction, plus $315 million in Section 108, $15 million of which is designated for the county. Because the $315 million is in the form of loan guarantees, it must be paid back. In addition to the federal dollars, commercial banks also committed $210 million (initial commitment), bringing the total to $640 million. The initial allocation plan envisioned the following types of expenditure:

- Micro-loans (done mostly through the intermediaries)
- Business loans
- Commercial real estate loans
- Commercial loan guarantees
- Loan loss reserves
- Venture capital
- Business and technical support
- Economic development grant

Although the limitations of this single policy solution (LACDB) has become apparent to many in Los Angeles, I will argue that, in fact, the entire basis of area-based anti-poverty programs is flawed. Little can be accomplished when poverty is viewed as a neighborhood problem rather than the fate of individuals whose collective predicament is the end result of a social process. Furthermore, I will argue that in a metropolitan area such as Los Angeles, the diverse population requires a wider definition of poverty, focused primarily on capacity building. As such, community development should receive the highest priority, and economic development efforts should be viewed as important elements of this process rather than its sole agenda.

## LOS ANGELES POVERTY AREAS AND THE EMERGENT SUPPLEMENTAL EMPOWERMENT ZONE

An important reason for selecting the 41 census tracts for the initial EZ application (and the current SEZ) was the federal formula for identifying EZs. Not unlike the poverty definition, which is defined at the national level and applies uniformly to all areas of the country, the poverty area cutoff line is not adjusted for various urban areas. Similar to Greene's (1991) findings for Los Angeles in 1980, an examination of the 1990 census reveals a dispersed poverty population. Of the 1,652 census tracts in the county, 413 can qualify as poverty areas (tracts with more than 20% poverty rate) (see Table 9.1). Because of this spatial dispersion of the poverty-stricken population, the number of people living in the so-called poverty areas who do not qualify as poor is substantially higher than the number of those who are classified as poor (see Table 9.2).

In other words, as in the rest of the country, close to two-thirds of the poor people in the Los Angeles area live in non-poor areas. Either the poverty definition for individuals is incorrect or the area-based categorization is completely irrelevant to this urban setting. Figure 9.2, which depicts poverty areas and identifies the location of the Supplemental Empowerment Zone, illustrates the level of exclusion in this geographic exercise. After all, only 81,264 poor persons out of the county's 1.3 million poor reside in this zone.

As Tables 9.3 and 9.4 indicate, the SEZ area is not necessarily representative of Los Angeles's urban poor, because only 1 out of 10 poverty tracts were included in this designation. The SEZ contains less than half of the tracts with a 40% or higher poverty rate (21 out of 50 in the county, 40 of which are located in the city) and a much smaller share of tracts with 30%-40% poverty rates.

The problem becomes further complicated when populations of the areas identified as part of the SEZ are analyzed. Tables 9.3 and 9.4 provide a demographic profile of the populations residing in the 41 selected EZ tracts. As illustrated, a number of structural problems begin to emerge.

First, despite the numerical decline of the population and the relatively stable number of African Americans, Latinos are emerging as a majority in the city of Los Angeles and soon will outnumber other groups in the county. According to the 1990 census, Latinos made up 37% of the county, 39% of the city, and 56% of the SEZ population. Considering the fact that only 40% of the SEZ residents speak English at home, it can be reasonably assumed that a significant portion of the Latinos are recent immigrants (arrived since 1980). Among the Latino residents of the SEZ and some of the Latino professionals involved in the SEZ designation process, there appears to be a feeling that the final designation favored African Americans more than Latinos. In a racially and ethnically charged atmosphere of competition, one unintended result of the SEZ designation will be further exacerbation of fragile relationships.

TABLE 9.1  Poverty Rates in Los Angeles Area Census Tracts, 1990

| Percentage of Residents Below the Poverty Line | Number of Tracts (Supplemental Empowerment Zone) | Percentage of Tracts (Supplemental Empowerment Zone) | Number of Tracts (City) | Percentage of Tracts (City) | Number of Tracts (County) | Percentage of Tracts (County) |
|---|---|---|---|---|---|---|
| 0 to 19.99 | 0 | 0 | 467 | 65.2 | 1,239 | 75.0 |
| 20 to 29.99 | 3 | 7.3 | 116 | 16.2 | 233 | 14.1 |
| 30 to 39.99 | 17 | 41.5 | 93 | 13.0 | 130 | 7.9 |
| 40 or higher | 21 | 51.2 | 40 | 5.6 | 50 | 3.0 |
| Total above 20% | 41 | 100 | 219 | 34.8 | 413 | 25.0 |

SOURCE: Data are from U.S. Department of Commerce (1993). Values were computed by the author.

**TABLE 9.2**  Percentage of Poverty Population Living in Poverty Areas

| | All Tracts in United States | Poverty Tracts in United States | Los Angeles County | Los Angeles County Poverty Areas | City of Los Angeles | Los Angeles Supplemental Empowerment Zone |
|---|---|---|---|---|---|---|
| Number of census tracts | 61,255 | 13,612 | 1,652 | 413 | 716 | 41 |
| Mean poverty rate (%) | | 32.7 | 14.0 | 30.4 | 17.2 | 40.2 |
| Total population below poverty line | 31,748,864 | 15,214,268 | 1,308,255 | 729,344 | 643,809 | 81,264 |
| Total population | 248,709,873 | 48,329,148 | 8,863,164 | 2,471,452 | 3,485,398 | 211,365 |
| Percentage of population in poverty | 12.8 | 31.5 | 14.8 | 29.5 | 18.5 | 38.4 |

SOURCE: Data are from U.S. Department of Commerce (1993). Values were computed by the author.

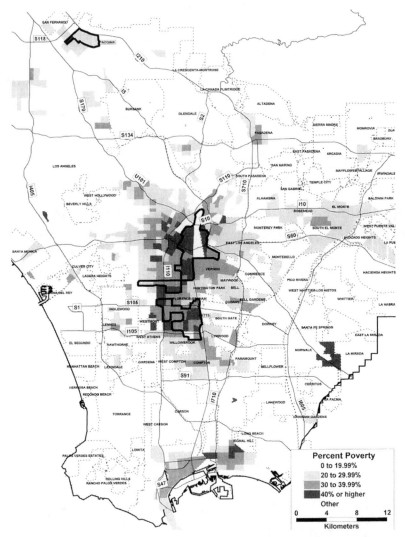

**Figure 9.2.** Geography of Poverty and Supplemental Empowerment Zones in Los Angeles County
SOURCE: The Urban Environment Initiative Data CD, Center for Spatial Analysis and Remote Sensing, California State University, Los Angeles, December 1997. Poverty data from U.S. Census, 1990.

Second, educational attainment of the population (25 years and older) in the city of Los Angeles is fairly representative of the county as a whole. In the SEZ, however, the lower levels of education are painfully

**TABLE 9.3** Population Profile of Empowerment Zone Residents in Los Angeles County, Compared to the Greater Metropolitan Area

| | Supplemental Empowerment Zone | City of Los Angeles | Los Angeles County |
|---|---|---|---|
| Total population | 211,365 | 3,485,398 | 8,863,164 |
| Families | 38,310 | 769,078 | 2,036,104 |
| Average number of persons per family | 5.52 | 4.53 | 4.35 |
| Non-Hispanic white | 6,841 | 1,305,647 | 3,634,722 |
| Non-Hispanic African American | 81,187 | 460,893 | 946,862 |
| Non-Hispanic Asian | 4,016 | 329,270 | 924,291 |
| Chinese | 578 | 69,795 | 248,415 |
| Filipino | 382 | 88,889 | 223,276 |
| Korean | 389 | 73,418 | 143,672 |
| Latino | 118,163 | 1,370,476 | 3,306,116 |
| Mexican American | 95,023 | 925,141 | 2,519,514 |
| Salvadoran American | 9,416 | 184,513 | 253,086 |
| Guatemalan American | 4,792 | 86,078 | 125,091 |
| English language spoken at home | 83,515 | 1,606,215 | 4,440,633 |
| Education[a]: < 9th grade | 38,227 | 401,207 | 853,988 |
| Education: 9th-12th grade, no degree | 27,493 | 318,232 | 788,825 |
| Education: High school degree | 20,796 | 419,318 | 1,134,608 |
| Education: Some college, no degree | 11,918 | 396,309 | 1,077,427 |
| Education: Associate degree | 3,819 | 144,377 | 402,932 |
| Education: Bachelor's degree | 2,764 | 318,802 | 793,556 |
| Education: Graduate+ | 1,669 | 181,659 | 429,886 |
| Per capita income ($)[b] | 5,769 | 16,188 | 16,149 |
| Per capita income ($): White | 5,860 | 22,191 | 20,531 |
| Per capita income ($): African American | 6,572 | 11,257 | 12,018 |
| Per capita income ($): Asian | 4,510 | 13,875 | 14,581 |
| Per capita income ($): Latino | 4,561 | 7,111 | 8,066 |
| Persons in poverty | 81,264 | 643,809 | 729,344 |

SOURCE: 1990 Census of Population and Housing.
a. Education statistics are provided for persons 25 years of age or older.
b. For the Supplemental Empowerment Zone, the income values reflect the mean of the 41 census tracts.

obvious. People with less than high school degrees make up only 33% of the population in the city, whereas in the SEZ, they make up 61.6% of the resident population. College-educated individuals make up less than 5% of the population in the SEZ, whereas the city has close to a 23% college-graduated population. Education and capacity building are essential in the SEZ.

Third, per capita income reflects the regional income disparity very clearly. Although the city of Los Angeles and the county are once again

**TABLE 9.4** Population Profile of Empowerment Zone Residents in Los Angeles County, Compared to the Greater Metropolitan Area

|  | Supplemental Empowerment Zone | City of Los Angeles | Los Angeles County |
|---|---|---|---|
| Employment participation rate (16 years or older) | 42% | 67% | 62% |
| Industry: Mining | 90 | 1,703 | 6,911 |
| Industry: Construction | 4,139 | 97,573 | 246,580 |
| Industry: Manufacturing nondurable goods | 9,125 | 130,004 | 307,002 |
| Industry: Manufacturing durable goods | 9,294 | 177,870 | 554,335 |
| Industry: Transportation | 2,925 | 63,675 | 186,041 |
| Industry: Communication and public utilities | 710 | 37,259 | 102,964 |
| Industry: Wholesale trade | 3,073 | 76,222 | 213,097 |
| Industry: Retail trade | 8,485 | 260,392 | 647,951 |
| Industry: Finance, insurance, and real estate | 2,260 | 135,214 | 327,998 |
| Industry: Business and repair services | 4,789 | 121,830 | 264,282 |
| Industry: Personal services | 2,749 | 79,177 | 156,643 |
| Industry: Entertainment, recreation | 739 | 71,223 | 130,529 |
| Industry: Health services | 3,281 | 118,360 | 302,332 |
| Industry: Educational services | 2,845 | 107,621 | 285,612 |
| Industry: Other professional services | 2,739 | 131,759 | 296,399 |
| Industry: Public administration | 1,687 | 37,442 | 120,901 |
| Occupied housing units | 52,494 | 1,217,405 | 2,989,552 |
| Living in the same house as 1985[a] | 90,091 (48%) | 1,485,904 (46%) | 3,837,105 (47%) |
| Home ownership | 27% | 37% | 46% |
| Renter occupied | 37,516 | 737,661 | 1,548,688 |
| Housing units built before 1939 | 13,323 | 226,380 | 424,273 |
| Housing units without complete kitchen facility | 4,160 | 25,305 | 44,637 |
| Housing units without complete plumbing | 2,215 | 12,763 | 22,932 |

SOURCE: 1990 Census of Population and Housing.
a. Computed for individuals 5 years of age and older.

fairly similar, as expected, the SEZ has drastically lower per capita income. This problem is reflected by the employment participation rate (which excludes the unemployed and "not in labor force" population 16 years and older) of the SEZ residents (i.e., 42%). This variable can be thought of as simulating the "working to nonworking population ratio," in which the smaller number of breadwinners generally produce less income.

Fourth, occupationally, the employed population 16 years and older in the county engages predominantly in retail trades and manufacturing of durable goods. This holds true in the city as well, but in the SEZ, manufacturing of durable and nondurable goods dominates the labor market. More than 30% of SEZ residents are employed in these occupations, and if policy in economic development continues to favor manufacturing, this sector might become the only major source of employment for the area. What the SEZ seems to lack is employment in the financial, real estate, and professional services.

Fifth, housing quality, as measured by incomplete kitchen facilities and plumbing, is indicative of the fact that Los Angeles as a whole is far from having the characteristics of an urban slum. Whereas in Los Angeles County and in the city less than 3% of the occupied housing units lack kitchen facilities, in the SEZ these numbers reach 8%. This is better than older East Coast cities, which have a much larger dilapidated housing infrastructure. Considering the fact that only about 25% of houses in the SEZ were built prior to 1939, the housing structure has been kept operational partially because of the fact that home ownership is significantly higher in Los Angeles's poverty areas than other regions of the country. Home ownership in the SEZ is about 27%, whereas the city of Los Angeles and the county boast 37% and 46% home ownership rates, respectively.

Sixth, one of the most striking aspects of the population is its high level of mobility. Despite the poverty status of the SEZ neighborhoods, only 48% of the population has lived in the same house for more than 5 years.

Although the social geography of the selected census tracts reveals them as poverty-stricken and therefore deserving of development attention, it should not be surprising that their location alone, in the area known as "South-Central Los Angeles," made them prime candidates for assistance. From a political standpoint, it was imperative that this urban area, which America has seen in flames twice in less than 30 years, receive some attention, especially because little changed after the Watts riots. It is ironic to note that "South-Central," a toponym so widely used by the public, is geographically elusive. This was a sobering point for the first leader of Rebuild LA, Peter Ueberroth, who was unable to answer completely the question, "Where exactly is South-Central?" In fact, very few people agree on where this urban region begins and ends. A few weeks after the 1992 civil unrest, the Los Angeles Times ("The Story of South Los Angeles," 1992) defined the area as falling primarily west of Alameda Street, south of the Santa Monica Freeway, west of Cren-

**Table 9.5**    South Central Los Angeles Demographic Data

|                                              | 1990     | 1980     | 1965     |
|----------------------------------------------|----------|----------|----------|
| Ethnicity                                    |          |          |          |
| White %                                      | 2.7      | 15.8     | 17.4     |
| African American %                           | 44.8     | 66.7     | 81.0     |
| Latino %                                     | 50.1     | 13.7     | NA       |
| Asian %                                      | 1.9      | 2.0      | NA       |
| Unemployed %                                 | 8.6      | 6.3      | 5.6      |
| Not in labor force %                         | 41.8     | 46.0     | 47.7     |
| Percentage of households receiving welfare % | 24.9     | 19.1     | NA       |
| Adjusted average welfare income              | $5,988   | $6,023   | NA       |
| Adjusted median household income             | $19,382  | $16,592  | $14,635  |

SOURCE: "The Story of South Los Angeles" (1992), p. A23.

shaw Boulevard, and North of Rosecrans Avenue. Whatever its boundaries, this elusive geography has changed little since 1965; however, its
population has. With 8.6% unemployment, 41.8% not in the labor force,
and 24.9% welfare usage in 1990, the area has seen little improvement
since 1965 (see Table 9.5). As indicated earlier, however, the significant
growth of the Latino population has also demographically modified
this region. By 1990, Latinos accounted for 50.1% of the population,
whereas in 1965 African Americans made up 81% of the residents.
Images from the 1992 civil unrest attest to the population shift and
ethnic makeup of this area.

Despite its significant Latino presence, however, "South-Central"
remains a largely African American space. Given that the EZ initiative
was made more urgent by the April, 1992, events, it would have been a
political faux pas to neglect America's quintessential inner-city neighborhood, an area defined through media-fanned mass hysteria and the
middle-class urban mythopoeia as the center of the urban apocalypse.
Although "South-Central" is no Harlem, its location in the capital of the
American symbol of consumption chills the otherwise sunny dreams of
Los Angeles boosters and celebrators of late capitalist nirvana.

## DEVELOPMENT EFFORTS AND THEIR FAILURE PATTERN

The Supplemental Empowerment Zone designation and its associated
dollars are merely a new chapter in this metropolitan area's attempt to

improve its blighted areas. Among its various strategies and urban renewal efforts, the Community Redevelopment Agency (CRA) and Rebuild LA are the best examples of two distinct eras in policy perspectives. One focused on housing and urban infrastructure improvement, and the other on economic development.

Probably one of the oldest community development efforts in the region, the CRA was established in 1948 as a result of the California Redevelopment Act of 1945. Through this act, it was hoped that cities and counties would rebuild their blighted areas and create a large number of jobs through the urban renewal projects. The CRA, however, became the champion of downtown development and assisted in creating the corporate center that exists today. Of the 210 projects completed between the late 1950s and 1990, 116 were in the central business district and the adjoining Bunker Hills area (Regalado, 1992). The remaining 94 projects were spread over the CRA's other 18 development areas. These included some of the poorest neighborhoods encompassing South-Central. This agency's focus on housing and physical planning rendered it less than useful to a population whose housing conditions were generally better than most minority neighborhoods in American industrial cities.

Inner-city Los Angeles began to witness a significant job loss in the years after the Watts riots and continued to experience a gradual departure of major manufacturing employers. What this population needed was poverty prevention, not urban renewal. The task of creating a more attractive downtown for the emerging financial center and the posh residential towers for its employees was masterfully performed at the cost of neglecting South-Central and all other poverty-stricken neighborhoods.

By the 1990s, we supposedly knew better. When the 1992 civil unrest occurred, America, especially the shaken-up Los Angelenos, jumped to help. Driven by the logic of the 1980s, however, it was envisioned that business, government, and the community would form an alliance to "bring prosperity to neglected communities by securing substantial outside private and public sector investments, . . . cutting through red tape; and through volunteerism" (Griego, 1997). This meant old-fashioned economic development, with a focus on job creation. Not surprisingly, because Rebuild LA was created as a response to the civil unrest, its target area was South-Central. The initial phase, which lasted through early 1994, was headed by Peter Ueberroth, who delivered the funding commitments necessary for the life of this program. Given that he and the other four cochairs were defined as outsiders, their efforts were seen as "top-down" and fragmented (Griego, 1997).

The second phase focused more on community. Under the leadership of Linda Griego, former deputy mayor of the city of Los Angeles for economic development, Rebuild LA began to establish an impressive presence in the community. By 1997, when Rebuild LA ended its 5-year proposed life and transferred its activities to LA PROSPER, a newly created economic development organization under the auspices of the Los Angeles Community College District, the level of investment in the "neglected areas" (i.e., South-Central) had approached $389 million. This translated to the creation of 17 new supermarkets and more than 1,000 jobs (Griego, 1997). In addition, because Rebuild LA was interested primarily in economic development, it expended a significant level of effort in researching the economic structure of the "neglected areas" and used the findings to formulate policies and programs around six major industries: manufacturing, biomedical technologies, ethnic food processing, textile and apparel production, the toy industry, and household furniture. In each of these areas, Rebuild LA moved beyond research to help develop steps for further economic growth of the targeted communities.

Despite its relative success, Rebuild LA's final report (Griego, 1997) contains three sobering remarks for those favoring place-based development programs as the answer to America's inner-city problems. First, through its research, Rebuild LA learned that in fact, "neglected communities" were not the economic wastelands everyone envisioned. They were found to house an impressive economic base of existing businesses and had great potential for further expansion. Second, large developments (e.g., suburban style supermarkets) do not always work in high-density inner-city neighborhoods, where large parcels of land are scarce and their cost is prohibitive. Third, shifts in the economy have a profound impact on inner cities, and no amount of incentives will make companies remain spatially bound. Downsizing (or right-sizing) will translate to net job loss, as was the case in Los Angeles during the early 1990s. As a more stable economy reappeared in 1995, Rebuild LA began to focus on small businesses, especially in the manufacturing sector, where the greatest need existed. Despite Rebuild LA's clear success in its attempted projects, the regional and even the target area impact of this program seems negligible. As thousands of jobs are lost to South-Central, Rebuild LA has replaced but a fraction. Although this does not detract from the organization's success, it does send the message that a more substantial regional approach might be necessary to show a statistically significant improvement in citizens' quality of life. After all, there are more poor people living outside the targeted areas in the county than there are inside those few census tracts. It is ironic that as the SEZ has copied Rebuild LA's spatial focus, its economic development style is even more laissez-faire and involves no policy direction. For that, we look to LA PROSPER to utilize the resources of the LACDB to advance the legacies of Rebuild LA.

## SUMMARY DISCUSSION AND CONCLUSIONS

Like previous area-based anti-poverty programs, the new Supplemental Empowerment Zone fails to incorporate any strategy for stabilizing poverty-stricken neighborhoods and as such creates just another layer of bureaucracy for disseminating funds. The large sum of money designated for the SEZ is targeted toward a single solution, business lending, without any focus on social services and capacity building. Policymakers hope that the city and county economic development programs and other social services will address this problem through their existing initiatives, including job training programs. This blind hope for the additive value of existing programs is indicative of the fact that poverty is being redefined simply as areas where people are unemployed because of a lack of investment by the private sector.

The early language of the SEZ highlights three distinct assumptions in the design of this initiative. First, it is implied that Empowerment Zone and Section 108 areas (census tracts with more than 20% poverty) have the same socioeconomic conditions. Second, the language implies that areas serving the Empowerment Zones are also fundable, and third, it implies that the LACDB approach can generate funding for social services, and better yet, further economic development of the area. Statistics discussed in this chapter cast some doubt on this philosophy, however, because the poverty-stricken population is well dispersed throughout Los Angeles.

The language regarding businesses serving the Empowerment Zone is rather intriguing to those familiar with the concept of Empowerment Zones, because it implies that in Los Angeles, a different interpretation is being offered. Here, it is believed that if a business is located or will locate within 1 mile of the EZs and 51% of its employees live in the EZs, it will qualify for assistance. Of course, there are no requirements for how long a company must maintain a 51% employment from the zone or what proportion of the wages, which determines the types of employment, should be received by the employees from the zone.

Given that the sole strategy in Los Angeles is establishing a lending institution and that Section 108 money cannot be invested in risky loans, the LACDB will benefit primarily the portion of the population whose poverty status is transitory and who are fully capable of profiting from some income boost. More than half the poverty areas' residents are new to their homes. Additionally, two-thirds of the population in the designated SEZ actually does not meet the definition for poverty; therefore, it can be concluded that an economic development focus will most likely lead to their eventual departure from the area, with minimal benefits for the remaining population. Based on the data discussed earlier, this segment consists mainly of the more permanent members of the community, whose poverty is most likely shaped by high dependency ratios, a disproportionate female to male ratio, and expenditure of a disproportionate amount of income for rent, not food. If these individuals are to be the target of SEZ effort, a more comprehensive program, which ties job training programs to actual well-paying jobs, must materialize. Otherwise, although bookkeeping might reveal that hundreds of jobs are created, the actual impact on the communities' overall improvement, even in terms of an absolute increase in the median household income, will be negligible.

Los Angeles, not unlike the rest of America, has begun its honeymoon period with the business community and has accepted its vision as the true image of success. Los Angeles has elected a business-oriented mayor, Richard Riordan, with business-oriented strategies. As a result, the city will once again have to discover the communities it left behind. These are the thousands of poverty-stricken, homeless, and chronically unemployed whose lives surely will not improve substantially as a result of the area-based Supplemental Empowerment Zones and Enterprise Communities. It is ironic that while private sector business initiatives such as Rebuild LA have failed to produce a significant change in Los Angeles' economically distressed neighborhoods, the Community Development Bank, with its loans but no integrated power or authority to deliver other community development efforts, hopes to do better.

The geography of a county with a dispersed and culturally diverse population, and with a multitude of racial and ethnic differences (real or perceived), should be forbidding to those policymakers who hope to develop a unified homogeneous solution for the development of its communities. Probably the most difficult barrier to overcome is the political fragmentation of this metropolitan area. Although the nine million residents may not be aware of their geographic location with reference to the spatial bureaucracies built around them, the county, the city, and the other 80-plus municipalities approach their development efforts as if they were isolated islands unaffected by neighboring communities. For example, academic literature is rich with debates over the deindustrialization of American cities and a number of scholars have thoroughly addressed the problems of Los Angeles's job flight (Grant, Oliver, & James, 1996; Ong & Valenzuela, 1996; Soja, 1989), but the governing body has failed to achieve a regional consensus for dealing with urban restructuring. Instead, economic development is expressed chiefly by competition and disjointed efforts, a pattern consistent with the observed political, cultural, and group fragmentation in the region.

Similarly, at the neighborhood level, little hope exists for some agreement on how the region should or could be developed. This is to be expected, because most community-based organizations have a very narrow geographic focus. The added dimension of cultural diversity also produces dilemmas. There is little evidence that any multicultural collaborative has been successful over the last 10 years. What remains is a landscape of poverty governed by layers of bureaucracy and political machines that cater to the emerging ethnarchy. In this climate, there is little hope that Latinos will achieve a convincing collaboration with African Americans or that Koreans will see eye to eye with either of the two. When one considers the hundreds of ethnic groups in Los Angeles, there remains little hope that old fashioned trickle-down economics will produce a level playing field for all parties.

Added to this problem is the national ineffectiveness of place-based anti-poverty urban policies. The history of Los Angeles is rich with such failure patterns. Although the CRA, Rebuild LA, and the recently arrived LACDB might be viewed as major urban efforts with significant problems, failures among smaller business and community-based initiatives also abound. The Watts Industrial Park, a light industrial complex developed in the aftermath of the Watts Riot, provides a good lesson in providing nonsustainable community/economic development. Lockheed, a large manufacturing corporation with a significant presence in the Southwest, located one of its sites in the park, as did several other large organizations, to provide accessible jobs to the community. The jobs were within walking distance of the workforce. The company established a child care center nearby so that workers could maintain their working status while feeling assured of their children's safety, and so they could remain closer to their families.

During the first year of operation, the large parking lot provided for commuters held only a few cars, because most workers got to work by walking or a short transit. During the second year, a few automobiles began to show up on the lot, until gradually, by the end of the third year, the lot was full. It was then that workers, as permitted under their union contract, began transferring to other Lockheed locations. At the same time, little attention was given to the underbelly of the urban infrastructure—schools, housing, commercial development, street lighting, and quality of life issues such as safety of the environment. There was no development of the community, and those people who could, chose to leave.

With the commitment of Lockheed, and a community development approach, success might have been more likely. Instead, the situation created a population flow, taking resources—human capital—out of the very community the park hoped to improve. The ones leaving were the individuals and families who had attained a better economic status through the availability of jobs at Lockheed, and those remaining behind were one step closer to becoming the late 20th century social residuum—"the underclass," in popular jargon. This is, in fact, the irony that most economic development advocates have failed to comprehend, and with that lack of comprehension, they have become agents for further social fragmentation and diminished hopes.

# CHAPTER 10

# MIAMI

*The Overtown Neighborhood—*
*A Generation of Revitalization*
*Strategies Gone Awry*

Dennis E. Gale

As American cities go, Miami is young. Incorporated in 1896, it is neither an old industrial city nor an antebellum southern community, but rather a 20th century Sunbelt city. Its subtropical climate, tourism-dependent economy, and orientation to the Caribbean and Latin American regions renders it difficult to compare with other American cities. It is a community with sharply defined racial and ethnic divisions, exemplified by neighborhoods such as Little Havana, Little Haiti, Liberty City, and Coconut Grove. None of these, however, is as severely distressed as Overtown, which lies just north of Miami's central business district (see Figure 10.1).

Although Miami was not settled until well after the Civil War and Emancipation, lynchings and beatings of African Americans at the hands of white mobs were documented by the early 20th century (Portes & Stepick, 1993, p. 80). By the 1930s, these atrocities had subsided, and practices such as restrictive covenants, racist rental policies,

159

**Figure 10.1.** Miami and Overtown
SOURCE: Prepared by Northern Ohio Data & Information Service, a member of the Ohio GIS-Network, The Urban Center, Levin College of Urban Affairs, Cleveland State University.

cross burnings, bombings, arson, and political disfranchisement were practiced by whites to segregate and control people of color. Blacks of Caribbean descent, as well as African Americans, were crowded into a few neighborhoods in apartheid-like enclaves. Of these, Overtown— then known as Colored Town—was for many years the largest and was considered Miami's African American business and cultural center (Mohl, 1983).

A visit to Overtown today reveals a community that is only a shadow of its former self. The massive interchange of I-95, I-395, and State Road 836 has disrupted the street grid and divided the neighborhood. A cave-like world exists under the raised highway, and homeless people, substance abusers, and others subsist in its forlorn shadows. Vacant lots and boarded-up buildings abound. About one-fourth of Overtown housing units were vacant in 1990. Two- and three-story unornamented concrete apartment buildings with flat roofs predominate. Built after

World War II, many were poorly constructed, and today most are ill-maintained. Less than 6% of dwellings were owner-occupied single-family units in 1990. Under the few remaining trees, knots of working-age adults loiter on vacant lots out of the hot sun. Beer cans and bottles are passed around, and strangers driving through the area are viewed with eyes either suspicious or glazed. Children and teenagers hang out in doorways even during school hours. In short, Overtown is a picture of severe distress. How did it reach this sad state?

First, I examine the influence of population migration on the political and economic life of South Florida. Second, I explore the politics of marginalization and its effects on African Americans in Miami. Third, I offer a brief assessment of recent population and social characteristics in Overtown. Finally, I review efforts at community revitalization and offer a critical analysis of outcomes to date.

## IMMIGRATION, POLITICS, AND ECONOMICS IN MIAMI AND DADE COUNTY

For most of the 20th century, Florida's growth has resulted primarily from migration from other states and Canada. Beginning in the 1970s, migration from other nations, particularly Cuba, rose sharply. Over the 1980s, however, the United States experienced the highest influx of immigrants since the 19th century, and Dade County ranked fourth behind Los Angeles, New York, and San Francisco in its growth resulting from people from abroad. It ranked only behind Los Angeles and New York in the number of Latino immigrants (more than 145,000) (Frey, 1995, p. 733).[1]

The University of Florida estimated that Dade County gained more than 173,000 Hispanics from 1990 to 1995, or almost 40% of Florida's new Hispanic residents.[2] The 1990 decennial census found that of Dade's 1.94 million people, 953,000 (49.2%) were Hispanics (largely Latinos). Indeed, Hispanics have been the most rapidly rising ethnic group in every decade at least since 1950 (Portes & Stepick, 1993, Table 8). By the year 2000, it is likely that Dade will contain two Hispanics for every white. This shift will be due not only to rapid rises in Hispanic population but also to declines among whites, at least since 1980. In 1990, whites in Dade numbered 586,000 (30.3%) and blacks (including black Hispanics) numbered 369,000 (19.5%). (Portes & Stepick, 1993, Table 8).

In the city of Miami, Hispanics composed an even larger share of the 1990 population (60%) and whites a smaller proportion (12%). Blacks (including black Hispanics) amounted to 26%, or a slightly larger share than their counterparts in Dade (Stowers & Vogel, 1994, pp. 64-66).

Among Latinos in Dade and Miami, Cubans and Cuban Americans are by far the largest nationality. Colombians, Nicaraguans, Peruvians, and Puerto Ricans compose smaller but significant subgroups. Yet even Latino cultures do not express the full complexity of Miami's ethnoracial dynamics. Since its founding, the city has had a modest number of Caribbean blacks, and Haitians have become the most numerous among these groups (Mohl, 1983a). By 1996, there were almost certainly more than 100,000 Haitians there, making Cubans, Haitians, and African Americans the most visible ethnoracial subgroups in South Florida. Today, Little Havana and Little Haiti, along with Overtown and Liberty City, exemplify the fragmented, segregated nature of ethnic and race relations in Miami.

## Politics

Prior to 1973, only non-Hispanic whites had been elected as mayors of Miami. Thereafter, two Latinos, Maurice Ferre (Puerto Rican) and Xavier Suarez (Cuban American), occupied that office, coinciding with a substantial rise in the Latino electorate in Miami (Stowers & Vogel, 1994, pp. 71-72). Steve Clark, a white, served as mayor until his death in 1996. In a special election, Joe Carollo, a Cuban American, succeeded Clark. Never in the history of Miami municipal government has an African American, or a black person of any background, won the mayoralty. African Americans and Latinos have served for years, however, on the City Commission. In 1993, Miami elected one African American, along with two Latinos and two whites, to that body (Stowers & Vogel, 1994, p. 72). The African American political presence is a tenuous thing in Miami.

At the Dade County level, one white—Steve Clark—served as county mayor from 1970 until 1992, when the office was abolished. Later reinstated, the mayor's office was won in the 1996 county election by Alex Penelas, a Cuban American. He garnered about 90% of votes cast by Dade Hispanics. His chief opponent, African American Arthur Teele, former commission chairman, won about the same percentage among county African American voters. White voters split about evenly for the two candidates. As a result, seven Latinos, three Whites, and three African Americans sat on the commission as of the end of 1996. Within the city and county governments, Latino and African American civil servants have significantly increased their numbers (Stowers & Vogel, 1994, pp. 75-80). Members of both groups have been appointed to high offices (Castro, 1992, p. 125; Warren, Corbett, & Stack, 1990, p. 165) and have been elected to the Florida legislature and the U.S. Congress. In the final analysis, however, whites and Latinos maintain definitive political hegemony over African Africans (Dunn & Stepick, 1992, p. 43; Warren et al., 1990, p. 165).

One recent study concluded that African Americans have suffered a "double marginalization" or "double subordination," first by whites and more recently by Latinos (Portes & Stepick, 1993). The national Civil Rights Movement of the 1950s and 1960s gave African Americans hope that equality of treatment in housing and employment, along with access to education, would soon follow.

Meanwhile, Cubans escaping Fidel Castro's Communist government trickled into Miami and began to compete with African Americans for low- and moderate-income housing and low-wage jobs, especially in the tourism industry. Cubans brought entrepreneurial talents that resulted in the start-up of many small businesses (Perez, 1992, p. 92). Unlike other immigrant groups in history, which rose in affluence and power by working within the larger local economy, Cubans tended toward establishment of "enclave economies" (Portes & Bach, 1985). In such networks, goods and services are purchased largely from other Cubans or Latinos, rather than from business owned by African Americans or whites. Sharing a common language, Cubans were much more likely to employ other Cubans or other Latinos than African Americans or whites in their businesses. As minorities, Latinos were able to qualify under federal, state, and local equal employment and affirmative action programs for government subsidized assistance such as federal Small Business Administration loans (Dunn & Stepick, 1992, p. 52; Portes & Stepick, 1993, p. 46). At the beginning of the 1980s, the Mariel refugee crisis brought more than 50,000 Cubans into South Florida, straining the region's capacity to provide for the newcomers.

African Americans took sharp issue with the anomaly of foreign minorities receiving government assistance and "getting ahead" while American natives continued to suffer from poor housing, unemployment, and racial discrimination (Stack & Warren, 1992, p. 167). Thus, a combination of political and economic strides made by the newcomers led to sometimes contentious and bitter relationships between Cuban Americans and African Americans. This relationship is compounded by the fact that many Cubans embrace conservative Republican values of self-reliance and anticommunism, and do not look kindly on perceived indolence and welfare dependency (Perez, 1992, pp. 94-97). Politically moderate whites have left Miami for the suburbs, with the 1990 white population being no more than one-fifth that of 1960 (Stowers & Vogel, 1994, p. 67).

If African Americans in Miami have had difficulty competing with Latinos and whites, the arrival of large numbers of Haitians over the past 20 years has raised a new concern. Like Cubans, Haitians tend to show a strong work ethic and bring a colorful and well-developed culture and language (French Creole) with them. They are centered in Little Haiti, which lies but a 10-minute drive to the north of Overtown. Haitians and African Americans, however, rarely mix socially (Stepick, 1992, p. 67), and there have been no attempts to build political coalitions based on black identity. As Haitians and African Americans struggle to achieve the American dream, many in the latter group have begun to suspect that soon they will be faced with "triple marginalization." Coming at the hands of other black people, however, a third level of subordination would pose a particularly bitter pill for many in Miami-Dade's African American population. (Portes & Stepick, 1993).

## DISCRIMINATION, SEGREGATION, AND INTERRACIAL MOB VIOLENCE

Frustration among Miami's African Americans predates the rise in immigration, however, and is rooted in a historic pattern of racially exclusionary practices. For example, residential patterns of racial segregation in Miami have been among the most extreme in the nation. As long ago as 1940, one scholar identified the city as the most segregated among 185 cities in the United States. In 1950, Miami was rated the second most segregated city nationally (Mohl, 1983, p. 69). Gradually, its relative segregation has diminished, although research at the University of Michigan found that in 1990 Miami was the 39th most segregated among 318 metropolitan areas (Borenstein, 1994, pp. 1A, 24A). A major influence on segregated residential patterns are the dynamics of the housing market. One study of 20 U.S. cities found that the incidence of unfavorable treatment of blacks in the rental of housing was highest in Cincinnati, Dayton, and Miami. Blacks in Miami were much more likely than Hispanics or whites to suffer discrimination in rental housing (Turner, Struyk, & Yinger, 1991, pp. 44, 51-52).

A measure of the divisiveness of race and ethnicity in Miami is the fact that the city was the site of no fewer than four significant outbreaks of interracial mob violence in a period of approximately two decades. The first episode broke out in Liberty City in 1968, at a time when hundreds of such tragedies occurred throughout the nation (Wikstrom, 1974). Three riots occurred in the 1980s (1980, 1982, 1989), a decade in which no other major North American city suffered even a single such tragedy (Gale, 1996, pp. 104-107). The 1980 riot, by far the worst, spread throughout a large area to the north and northwest of Overtown. Marked by especially vicious African American attacks against whites, the event resulted in 18 deaths, more than 400 injuries, approximately 1,100 arrests, and damage estimated at about $804 million (Dunn & Stepick, 1992, pp. 43-45; Gale, 1996, pp. 104-107). A study in 1985 found that the federal government spent $70.6 million in response to the problems raised by the 1980 riot, of which at least $43.2 million was directly targeted at the neighborhoods affected by the violence (U.S. General Accounting Office, 1985, Table 1).

Rioting erupted in Overtown in 1982, leaving 26 injured and 43 arrested. In 1989, violence broke out again as Miami prepared to host the Super Bowl. The result was 1 death, 6 shootings, 27 buildings burned, and more than 400 suspects arrested. The event ignited in Overtown and later spread to Liberty City. In all three of the 1980s disturbances, questionable actions by white or Latino police officers and violent response by blacks were at issue (Dunn & Stepick, 1992, pp. 44-45).

## HISTORICAL PATTERNS, URBAN RENEWAL, AND HIGHWAYS

By the 1950s, Overtown had become a densely populated community of low-, moderate-, and middle-income households. Some Overtown landlords, who were opposed to public housing, began removing small frame dwellings after World War II and replacing them with two- and three-story, privately owned concrete apartment buildings (Mohl, 1993, p. 127). Later, opposition declined, and public housing and other subsidized units were located in Overtown. Today, nearly all the housing in the neighborhood is government owned, and at least three-fourths of multifamily units receive government subsidies (City of Miami Planning, Building and Zoning Department, 1993).

Federal urban renewal brought three projects into the area, although these were small in scale and physically disparate from one another. The federal interstate highway program, on the other hand, has had the most enduring negative impact on Overtown and its people (Mohl, 1993, pp. 139-140). Beginning in the mid-1960s, an interchange linking I-95, I-395, and State Road 836 was constructed in Overtown, wiping out 87 acres composed mostly of homes and retail shops (Mohl, 1993, p. 134). Although the federal interstate highway program eventually would include assistance to relocate those displaced by the highway right-of-way, such services did not exist during the early stages of land acquisition in Miami. Thus, many hundreds of African American households were forced out of Overtown with little or no relocation assistance (Mohl, 1993, pp. 119-132). In some cases, families dislocated by the new highway moved to public housing projects in Liberty City and elsewhere (Mohl, 1993, p. 124). In others, families moved to other parts of Miami or suburban Dade County for better housing and public schools.

## POPULATION AND SOCIAL CHARACTERISTICS IN OVERTOWN

From a densely populated neighborhood of 40,000 people in 1950, Overtown had become an area housing about 12,000 residents by 1990 (Mohl, 1993, p. 127).[3] Fully 83% of Overtowners were African Americans, compared with 25% of the population citywide (U.S. Department of Commerce, 1993). Just 3% of Overtown residents were white non-Latinos, and only 12% were white or black Latinos. Even the small proportions of whites and Latinos, however, tend to live on the geographic fringes of Overtown.

If racial concentration characterizes Overtown, so also does poverty. Fully half (51%) of households there and one-third (32%) citywide earned less than $10,000 annually in 1990. About equal shares of Overtown (35%) and Miami (33%) households earned between $10,000 and $24,999 annually. Only 15% of Overtowners and 35% of Miami residents earned $25,000 or more. Moreover, the average household size in Overtown was 3.14 people, versus 2.75 in the city. Thus, Overtown household incomes have to be stretched farther. Fifty-four percent of Overtown's 1990 population lived at or below the poverty level, compared with 31% in the city overall. Not only did the neighborhood have the highest poverty rate, it also had the highest proportion of female-headed households in Miami (City of Miami Planning, Building and Zoning Department, 1993).

In part, these socioeconomic disparities are explained by factors such as educational attainment among persons aged 25 years or older, but education alone does not explain socioeconomic variations. The unemployment rate in 1994 in Overtown was 23%; in Miami, it was 11% (U.S. Department of Labor, 1994). Unlike most of the cities of the North, Miami has never been an industrial manufacturing center. Thus, union-based, blue-collar, skilled jobs are not widely available. It is the tourism industry and jobs such as waiter, bellboy, hotel maid, gardener, security officer, and cook that traditionally have employed many minorities in South Florida. These low- and moderate-wage jobs increasingly have gone to immigrants, especially those from South America and the Caribbean islands.

Overtown's high rate of distress is reflected in its relatively high crime rates. Although the community has about 3% of Miami's population, in 1995 it sustained 8% of Miami's murders, 6% of rapes, 7% of robberies, and 10% of aggravated assaults (City of Miami Police Department, 1996).

## Ethnoracial Politics in Overtown and Miami

An examination of the 1996 Miami election results sheds more light on the difficulties that African Americans, including those in Overtown, face in gaining power. Carollo, a former member of the City Commission, won with three-fourths (16,556) of the votes cast. The runner-up and most of the other contenders also were Latinos. Placing third in the city, however, was C. C. Reed, an African American from Overtown. With higher name recognition than Reed, Carollo succeeded in convincing many in the Overtown neighborhood that in the larger electoral arena of city politics, a vote for him was an unwasted vote. Although having a more contentious political past than Reed, Carollo won easily, even in the overwhelmingly African American Overtown community.

Hurting Reed and other African American candidates who have run for office is the historically low voter turnout in African American neighborhoods such as Overtown. Fewer than one-fifth of registered Miami voters voted in the July, 1996, election, but 8.4% of those in Overtown voted (City of Miami, 1996). Citywide, Latinos had by far the highest voter turnout. Although both city and county elections are nonpartisan, the higher Republican voter participation rate demonstrates the amplitude of conservatism toward social issues held by the predominantly Cuban American electorate.

For 30 years, it was of some comfort to Miami's African Americans that one of their own was among the five members of the City Commission. In 1996, however, even this position was lost.[4] For the first time since 1966, the City Commission was composed exclusively of white and Latino members. The political quandary of African Americans has been termed the "Miami syndrome." It is underscored by "high levels of black frustration" compounded by a "stagnant political system" and "rising expectations" among African Americans; Miami's high incidence of mob violence thus is "symptomatic" of this extreme level of futility (Stack & Warren, 1992, p. 167).

## STIMULATING COMMUNITY REINVESTMENT

Among the indigenous institutions remaining in Overtown are a few churches and their community development corporations. Two—St. John's Episcopal Church and Bethel AME—focus on housing production. St. John's Community Development Corporation (CDC), for example, has developed or renovated 35 units of rental housing near the church. These units, however, are old and small, and their renovation has had relatively little impact on the neighborhood (personal communication, Donald Benjamin Interview, June 24, 1996). CDCs associated with St. Agnes and Mount Zion churches concentrate on housing management in Overtown (personal communication, Gregory Gay, June 24, 1996).

Other efforts have been directed toward small business development and enhancing retail opportunities in Overtown. The 30,000-square-foot Overtown Shopping Center, built by the city for $2.5 million as an "appeasement project," was opened in 1983 in response to earlier riots (personal communication, Gregory Gay, June 24, 1996). The project failed to attract enough commercial tenants and later was renovated by the city using $200,000 in federal funds. By mid-1996, the center still was not fully occupied, and at least one tenant, a supermarket, was in arrears in rent and utility bills. Several management companies have failed to operate it successfully, and the city now runs it (personal communication, Gregory Gay, June 24, 1996).

Banking also illustrates the difficulties of attracting new investment in Overtown. For years, no bank branch existed in Overtown. After complaints surfaced under the federal Community Reinvestment Act, several mainstream bank officials proposed creating a shared risk bank. Executives of the city's only African American–owned bank called the measure "racist" and said it would dilute the market for banks such as theirs. The mainstream banks hastily withdrew their proposal, although an African American bank branch has not yet been opened. In 1996, however, Republic National Bank, a Latino-affiliated institution, established a bank office at the beleaguered Overtown Shopping Center.

Efforts are under way to renovate the N.E. 9th Street corridor. It connects upmarket Biscayne Boulevard and the Atlantic coast on the east to the Overtown core and the Overtown Folklife Village in the center of the neighborhood. Some observers feel, however, that it was a strategic error to revitalize Overtown's center first. Miami's former planning director argues that neither public investment alone nor the limited resources of Overtown's residents and businesses can hope to revitalize the neighborhood successfully (personal communication, Sergio Rodriguez, July 19, 1996). Because Overtown is walled off from much of the city by highways, its only hope is to draw investment to the center of the community from downtown and from affluent Biscayne Boulevard and the coastal corridor to the east. Thus, Rodriguez argued, initial investments near Biscayne Boulevard could have lured affluent shoppers and home seekers and anchored a gradual spreading of redevelopment and rehabilitation westward, block by block, toward the center of Overtown. Investing in Overtown's center first, he noted, may have been politically expedient, but it was not good planning. Whatever the case, to date the city has been plagued by complaints of drug peddling, prostitution, loitering, and other crimes along the N.E. 9th Street Pedestrian Mall.

In the late 1980s, the Miami Arena, home of the Miami Heat, was built on the southeastern edge of Overtown, linking to the central business district. A Metrorail transit station, vast parking lots, low-rise condominiums, and apartment towers were located nearby. Supporters argued that this massive investment would create significant employment and housing for Overtown residents, although there is little evidence today of such outcomes. Ironically, a larger arena for the Heat is being built outside Overtown, thus putting to rest the myth of Overtown economic revitalization resulting from professional sports facilities.

After the 1989 riot, Overtown residents, the Miami-Dade County Commission, and the late James Rouse's Enterprise Foundation created Overtown Neighborhood Partnerships. A multifaceted 5-year plan was prepared to improve the residents' quality of life, focusing on home management and life skills. The centerpiece was a program to help Overtown residents earn a general education diploma or community college degree. After 5 years of operation, the community college terminated its support, and the city now struggles to maintain initiatives such as community gardening, open air markets, and a revived Overtown merchants organization (personal communication, Gregory Gay, June 24, 1996).

Nearly $19.4 million of Community Development Block Grant (CDBG) funds were spent from 1975 to 1993 in Overtown, mostly on housing rehabilitation, economic development, land acquisition, and building expansion (City of Miami Planning, Building and Zoning Department, 1993, pp. 47-49). On a per capita basis, troubled African American and Latino neighborhoods have received considerably less. Although federal criteria determine the basis for CDBG distribution, the large disparities among neighborhoods and the persistence of Overtown's problems had slowly eroded residual sympathies among some citizens and public officials.

In December, 1994, Miami-Dade was one of 73 urban communities nationally to receive a designation under the federal Empowerment Zones/Enterprise Communities (EZ/EC) program. Little progress has been made in operationalizing the EC. In Miami-Dade, the federal subsidy of $3 million was too small to have much effect in such a large EC. Matching funds were pledged by Metro-Dade but were delayed, as was a $5 million loan commitment by local banks (personal communication, Frank Casteneda, July 16, 1996).

Funding is only part of the problem, however. The shape and size of the EC illustrates the sometimes crippling role that ethnoracial politics and the "Miami syndrome" play in Miami-Dade. The EC stretches about 4.5 miles from the Miami International Airport on the west to Biscayne Boulevard and the Port of Miami on the East, with Overtown, parts of several other neighborhoods, and the downtown included. Boundaries were drawn so as to give physically disparate, highly segregated, and racially and ethnically divided communities a part in the EC program. Homestead and Florida City, more than 20 miles to the south of Miami and severely damaged in Hurricane Andrew in 1992, also were included. It was thought that their inclusion would strengthen Dade's chances of receiving federal designation. To achieve a consensus on which to base the original application to Washington, Metro-Dade created a target area that is essentially unworkable.

Efforts to stimulate job development for zone residents at Miami International Airport or other employment centers within the zone grapple with union hiring rules, sluggish employment growth, and the demand for skills largely unavailable among zone residents. The county has been trying to establish an "Enterprise Community Council" to bring consensus among the many factions involved, but county and city officials have had problems agreeing on who should sit on the body. Also planned was a one-stop center near Overtown where small minority businesses could secure loans and technical assistance for locating or expanding in the EC (personal communication, Richard Towber, September 25, 1996).[5]

## CONCLUSION: OF POWER, PURPOSE, AND PERSISTENCE

Most African Americans in Miami, and especially those living in Overtown, continue to struggle with the Miami syndrome. They are not powerless; rather, their power is tightly circumscribed by Latinos and whites and in the future may be further subordinated by Haitian advances. African Americans are able to elect their own to city and county councils but not to the city or county mayors' offices. An Urban Institute study concluded that African Americans are unlikely ever to wield enough influence to achieve the level of resource redistribution they need (Stowers & Vogel, 1994, p. 81). A kind of political stalemate has resulted in which their power rarely advances and nearly always threatens to recede.

Worst affected by this pattern is Overtown. It has been left behind by most African Americans, who live in Liberty City or other parts of Miami and Dade. Its electorate has shrunk, thus reducing its political influence. Over the past 30 years, plan after plan, policy after policy, and project after project, has been proposed in the neighborhood. Many have been implemented. At best, these efforts appear to have bought time for Overtown residents, businesses, and institutions. At worst, they may have only contributed to the neighborhood's decline. Aside from the issues of ethnoracial power distribution in Miami, Overtown's difficulties lie in the absence of a viable and feasible vision for its future. Four models have thus far been employed by planners.

## Model 1: Public Housing and Urban Renewal

In this model, direction came largely from federal agencies. It featured clearance of blighted structures and construction of large numbers of public and subsidized housing units, mostly in multifamily projects. This helped to ensure that Overtown would become permanently identified as a mostly low-income, subsidized community. Many families were pushed out by highway construction and pulled out by open housing laws and policies, resettling in Liberty City and other Miami and Dade neighborhoods.

## Model 2: Indigenous Community Development

In this model, direction came largely from self-designed programs, projects, and plans, with financing from federal sources such as Community Development Block Grants. The model emphasized community self-determination through small-scale housing construction and rehabilitation, small business development, public facilities improvements, and social services provision by a combination of CDCs, municipal agencies, philanthropies, and private businesses. So far, this strategy has brought a degree of stability and spot improvements, but progress is very slow, and the net impacts are modest at best.

## Model 3: Historic Restoration and Cultural Pride

Development under this model is also largely internally directed and also largely federally financed. It focuses on restoration of remaining older structures linked to neighborhood cultural heritage. Primarily spurred by civic elites in Overtown, the goals are to renew the community's historic identity, create pride in its African and African American origins, and stimulate tourism and reinvestment.

## Model 4: Megastructures and Economic Spillover

Direction in this model came largely from city and county agencies and private capital. The model featured construction of Miami Arena, the Metrorail station, and associated middle-income apartment towers under public/private partnership. The intent was to stimulate secondary investment in Overtown and create jobs for its residents.

Model 1, of course, has long been discredited. Model 4 has done little to create jobs or induce new construction of benefit to Overtown residents. (With relocation of Miami Heat games to a new stadium on Biscayne Bay in Miami, this approach is doomed.) Models 2 and 3 continue to be applied but depend primarily on federal programs. Because subsidies such as Community Development Block Grants, SBA loans, and Empowerment Communities tax concessions are very limited, the pace of Models 2 and 3 will continue to be very sluggish. Both are intended to attract subsequent private investment in Overtown but have had only modest impacts so far. Both models focus largely on the needs and interests of existing residences and businesses in the neighborhood.

There appears, however, to be a disjuncture between the interests of capital in Miami and the willingness of the neighborhood's civic elite to accommodate alternative futures. Neighborhoods such as Coconut Grove, Little Havana, and Little Haiti retain strong, distinctive identities based in part on their cultural roots. Nearby Miami Beach (South Beach) has the largest concentration of Art Deco architecture in the nation, thus reinforcing its identity as a nightlife and cafe society enclave. Model 3 represents an attempt, primarily by Overtown's civic elite, to create or revive its own identity. Although preserving neighborhood culture is important, neither historic African American nor African themes appear to have the appeal necessary to draw middle-class people to live in Overtown nor tourists to visit there in significant numbers. Some observers argue that Overtown's "Old Guard" resists change and that those with alternative ideas about the neighborhood's future are given short shrift (personal communication, Gregory Gay, June 24, 1996). In Overtown's case, many of those who once created and maintained its cultural identity and uniqueness in Miami are deceased or have moved away. Most of the businesses and institutions no longer exist. Thus, the status to which these two models are attempting to revitalize Overtown is at best incomplete and at worst infeasible. To what future status, then, should Overtown be directed?

## A FUTURE FOR OVERTOWN?

A study and plan prepared for St. John's CDC in 1988 concluded that "The long battle to stem the destruction of Overtown as a viable black residential and cultural community has been lost" (Florida Center for Urban Design and Research, 1988, p. 10). It recommended that the community be revitalized for a mixture of racial, ethnic, and income groups capitalizing on nearby employment centers and public transit linkages. Under the proposal, most indigenous residents and businesses could remain and benefit, although the strongly centered historic African American identity sought in many Overtown proposals and projects was deemphasized. A later study by the same consultant provided detailed proposals for revitalizing the northern and western sections of the neighborhood by establishing a tax increment financing district (Florida Center for Urban Design and Research, 1993). Alternative estimates of needed investments ranged from $76.9 to $79 million (in 1993 dollars). More recently, St. John's CDC has produced a 3-year plan, drawing on these earlier studies and on views expressed at Overtown community meetings. The plan emphasizes benefits to existing Overtown residents and merchants but sadly downplays earlier recommendations for substantial middle-class residential reinvestment from outside the neighborhood (St. John's Community Development Corporation, Inc., 1996).

To this author, the most promising future for Overtown is to capitalize on its adjacency to downtown, Biscayne Boulevard, and Biscayne Bay; on its transit linkages to the city via its Metrorail station; and on its key employers, such as Jackson Memorial Hospital and the Omni Hotel complex. The vision for "Model 5" would be a truly multiethnic, multiracial neighborhood intended primarily for younger singles and childless couples. Additionally, empty nesters might be attracted as some households in suburban Dade seek convenient in-town living. A mixture of single-family structures, townhouses, and moderate-density apartments and condominiums, together with attractive neighborhood shopping and entertainment opportunities, would be planned. Higher-density housing might be appropriate on the eastern and southern perimeters of Overtown, but structures in the center would maintain, perhaps, two- and three-story configurations. Offices, artists' studios, retail shops, and incubator facilities for start-up businesses could provide an employment base. Health-related firms could take advantage of the medical community nearby. The southeastern edge of Overtown could accommodate some degree of downtown expansion, as long as a sharply defined edge and buffers between it and the residential community to the north were maintained.

Realistically, there is little likelihood that these newcomers would want to live near I-95, I-395, State Road 836, or several existing public and subsidized housing projects. A transitional belt of moderate-rent or mixed private market and subsidized units, however, could act as a transitional buffer. Historic structures such as the Lyric Theater and Dorsey House could be integrated into the new community and serve as cultural and historic assets. In the final analysis, though, existing residents of Overtown will have to become part of a more broadly conceived community than that proposed by current leaders. It is time to move forward.

## NOTES

1. Dade County encompasses the entire city of Miami, plus 25 other communities. It is a reform-style metropolitan government and shares power with these municipalities.

Both Miami and Dade County have council-manager forms of government, Dade with nine elected commissioners and Miami with five. Each has an elected mayor who is a member of the respective council. Elections are nonpartisan, and candidates run at large. Dade, commonly called "Metro" or "Metro-Dade," oversees services such as mass transit, public health, parks, and recreation. Miami is responsible for police, establishing tax rates, and zoning and planning, among other powers (Stack & Warren, 1992, p. 162).

2. Distinctions between and among race and ethnic groups in Miami-Dade are very complex and always imperfect. Here I distinguish between blacks as African Americans or Caribbean blacks (including Haitians). I also distinguish between Hispanics (all people of Spanish descent) and Latinos (Hispanics of Latin American descent, including Cubans). Where relevant, I distinguish between Latino whites and Latino blacks. Whites include people of European lineage but, unless otherwise noted, exclude Hispanics and blacks.

3. The city of Miami defines Census Tracts 30.10 (Block Groups 2 and 3), 31, 34, and 36.01 (Block Groups 1 and 2) collectively as the Overtown neighborhood.

4. In 1996, a Federal Bureau of Investigation corruption investigation led to the discovery of a $68 million city budget shortfall, fully 20% of Miami's annual budget. Much of the deficit was due to routine payoffs to political friends of city officials, mismanagement of funds, and poor collection of rents on city-owned properties leased by political supporters.

5. In January 1999, the U.S. Department of Housing and Urban Development upgraded Miami's Enterprise Community to an Empowerment Zone. Up to $500 million has been pledged in the form of loans and the incentives for job creation.

# 11

# New York

*Challenges Facing*
*Neighborhoods in Distress*

Thomas Angotti

## INTRODUCTION

Among the more than 100 neighborhoods that make up New York
City's diverse mosaic, many have achieved national recognition for
their extreme levels of abandonment, poverty, and crime. Legendary are
the stories of the South Bronx, Harlem, and Red Hook, for example,
which conjure up stereotypical images of devastation and ruin. These
distressed neighborhoods are only part of the New York story, which
also includes the elite enclaves of Park Avenue, Sutton Place, and
Riverdale, among the most prestigious addresses in the world, and a
large group of middle-income, working-class communities spread
throughout the city's five boroughs.

Visitors to New York's distressed neighborhoods today might be
surprised to see some of the changes under way. Significant portions of
the South Bronx have been rebuilt, population levels have stabilized,

and in some quarters there is a distinct sense of optimism. Conditions in most of the city's distressed neighborhoods have changed considerably since the devastating 1970s, when abandonment, arson, and destruction were under way, claiming block after block of what once were solid working-class neighborhoods. President Bill Clinton touted these changes in a December, 1997, visit to the South Bronx.

The rate of abandonment has slowed dramatically, and there is now evidence of new economic growth and housing development where previous abandonment took its toll. Much of the city's once-large stock of 150,000 housing units acquired through tax foreclosures has been renovated. Crime has gone down almost everywhere. The whole range of retailing activity, from street vendors to superstore outlets, has increased, making moribund commercial strips into lively streetscapes. Outwardly, things appear to be looking up.

The condition of relative poverty, however, remains and has worsened. Steady growth has brought even greater benefits to middle- and upper-income neighborhoods and households, expanding the gap between rich and poor. Much of the housing built on vacant lots is low-density, owner-occupied housing serving a distinctly middle-income population; except for some transitional and special needs housing, very little new housing has been built for low-income people. The city has renovated most of its vacant and occupied buildings for low- and moderate-income tenants, but conditions in many of these buildings are below standard, and the tenants remain among the poorest in the city. Crime is down everywhere in the city, but it still remains higher in poor neighborhoods, and in some categories, such as rape, there has been little change. New York City's unemployment rate is one of the highest among large cities; between 1983 and 1992, there were 133,000 fewer private sector jobs despite the relative stability in the population level (Department of City Planning, 1993, p. 27; see also Manhattan Borough President, 1995). In sum, there has been growth and development, but without equity. New York City still has a disproportionate share of areawide poverty.

About one-fourth of the population in the city lives below the poverty line, double the rate in the state of New York. In 1995, 18% of the population—well more than one million people—received public assistance or Social Security Insurance (SSI), over twice the U.S. rate. Female-headed households make up 17% of the population and 43% of the poverty population (Zimmerman & Li, 1995, pp. 7-9). The poverty population in the city has expanded significantly since 1975. The population receiving public assistance is concentrated in several areas: Harlem and Washington Heights in Manhattan; the South Bronx; and Bedford-Stuyvesant, Bushwick, and East Flatbush in Brooklyn.

The divide between wealth and poverty strongly overlaps with racial divisions. Although African Americans make up 26% of the population, they are 36% of the poverty population; Latinos are 24% of the total population and 39% of the poverty population. As a result, almost all African American and Latino neighborhoods have higher proportions of poverty than other areas.

Race and ethnicity remain the defining elements of neighborhoods in New York. In a city that is only 40% white, most African Americans still live in African American neighborhoods. For a city with perhaps the most ethnically diverse population in the world, New York is not a melting pot but a city of separate ethnic and racial enclaves, from Bensonhurst to Bedford-Stuyvesant. This pattern of segregation is continually reproduced as new immigrants settle in separate enclaves. The pattern is in part a product of durable family and ethnic networks among old and new immigrant groups. Between 1990 and 1994, more than half a million new documented immigrants came to New York. The largest group of immigrants is from the Dominican Republic, and more than half of them settled in the Dominican neighborhood of Washington Heights (Department of City Planning, 1996).

The segregation of neighborhoods is also a reflection of the durability of racial and ethnic discrimination. European immigrant groups have, over time, been gradually assimilated into multi-ethnic white neighborhoods and, most important, the suburbs, but enduring racism has prevented African Americans and other nonwhite immigrant groups from moving out of their segregated enclaves. Thus, New York City has about 40% of the New York metropolitan region's total population, but 90% of the region's African Americans live in the city.

## THE ROLE OF GOVERNMENT

Since the 1950s, many federal, state, and local government policies have focused on distressed neighborhoods. Government, in tandem with the real estate market, contributed to both the decline and redevelopment of New York City's neighborhoods. Its intervention has been contradictory, first aggressively displacing poor communities with urban renewal and redevelopment, then engaging in mildly benevolent policies that temporarily stabilized and improved parts of communities.

Federal, state, and local government in many ways helped produce severely distressed neighborhoods in the period from the 1950s to the 1970s—what can be called "Act One" of New York City's urban policy drama. The government programs from this period included the federal and local urban renewal and highway programs, which encouraged the movement of industry and labor to the suburbs and the redevelopment of land in central city neighborhoods for commercial and luxury high-rise buildings. Federal policy was the key catalyst to urban change during this period, though it responded to the needs of local real estate and industrial interests. In New York City, urban renewal helped remake the face of Manhattan's West Side, for example, as well as the neighborhoods near the budding business districts in the boroughs of Brooklyn, Queens, and the Bronx, where land values were rising.

The legacy from this era includes neighborhoods split apart and destroyed by highways that scar the assorted street grids that have given New York City its lively street life. There also remain pockets of blight in unfinished urban renewal areas scattered around the city. Without urban renewal, there would not have been public housing, so part of this legacy is the 185,000 units of public housing making up the largest local housing authority in the country (with one of the best track records for maintenance). This housing now accounts for some 5% of the city's housing stock, but it is disproportionately concentrated in low-income neighborhoods.

In addition, New York City was one of the only cities in the country to retain World War II era rent controls, thanks to a militant tenant movement able to win its perennial battles with suburban and upstate legislators. Until the 1970s, when the local real estate lobby joined with the suburban legislators to start the process of decontrol, the majority of rental units in the city were rent-regulated. Housing, therefore, was one of the main elements—along with home relief, municipal hospitals, and free public education up through the university level—that made up a safety net for poor households in the city. Federal and state assistance was critical in building this safety net, but New York City led the way in local efforts.

This safety net began to come apart in the 1970s. "Act Two" in the urban policy drama began in the 1970s, when the federal government started its withdrawal from assistance to cities. Antidisplacement struggles, often tied with the Civil Rights Movement (because neighborhoods with people of color were more likely to be the victims of displacement), contributed to the demise of large-scale urban renewal and highway programs. Funds for new public housing construction no longer were available.

A massive wave of property abandonment began in the 1970s, spurred by the movement of industry to the suburbs and abroad, bank redlining, landlord disinvestment, arson, and the city government's policies of "benign neglect" and "planned shrinkage" of poor neighborhoods (Homefront, 1977). The city became the owner of thousands of buildings and vacant lots whose private owners walked away. The city was slow to take title to abandoned properties. After taking title, the preferred policy has always been to sell land and buildings back to private owners as soon as possible, even if the new owners were speculators who would sit on the property and wait for land values to go up. Tenant and housing advocates and a vibrant squatters movement made it difficult for city government to carry out its intentions of clearing, demolishing, or reselling the buildings it had taken on tax liens. When massive homelessness occurred in the 1980s, partially as a result of the widespread abandonment, renovation of city-owned vacant buildings in low-income neighborhoods became the government's preferred solution, especially because city officials refused to face down the resistance to the construction of homeless housing in middle- and upper-income neighborhoods.

Community development corporations with roots in the earlier antidisplacement struggles and an interest in stabilizing and revitalizing their neighborhoods became partners with local government in efforts to rehabilitate abandoned housing. Prodded by these neighborhood organizations, New York City embarked on the largest and most successful municipal housing rehabilitation program in the nation. Over the last 10 years, the city has rehabilitated or constructed more than 125,000 units of housing as part of a $5.1 billion 10-year plan. In recent years, the city's housing agency actually built more housing than the private sector. Today, most of the city's vacant tax-foreclosed buildings have been rehabilitated for low- and moderate-income tenants and transferred to nonprofit or cooperative ownership. There are still more than 40,000 units of occupied city-owned units in 6,000 tax-foreclosed buildings (Department of City Planning, 1993, Figure 7-6).

Government policy in Act Two encouraged not massive redevelopment but a gradual process of upgrading and gentrification at the margins of many distressed neighborhoods. This has had mixed and contradictory results. It slowed the pace of displacement in some neighborhoods and quickened it in others.

Over the last two decades, local tax incentives for renovation of privately owned rental housing and coop conversion spurred substantial development in middle- and upper-income neighborhoods. Tax abatements on housing rehabilitation may have postponed the gentrification process by temporarily lowering housing costs, but in the long run they encouraged it; over a period of 15 years, renovated units gradually reverted to full tax payments, thereby ensuring increased rent charges.

Since the 1970s, community-based development organizations, born of movements to preserve neighborhoods, have managed to establish an important niche for themselves in housing and economic development. This has been possible to sustain only because of support from government. These organizations flourish because of the city government's efforts, supported by its bondholders and real estate backers, to rid itself of this massive stock of abandoned buildings. The city's most powerful and influential real estate owners generally are not interested in buildings with poor tenants or land in distressed neighborhoods. Thus, officials tend to see nonprofit organizations as the vehicle that will allow them to privatize the municipal housing stock and cut expenditures of local revenues and federal assistance for housing maintenance. Housing production by the nonprofit sector also relies heavily on the federal programs that have survived from the 1970s, especially Community Development Block Grants, Section 8 certificates, and syndication tax benefits through the Low-Income Housing Tax Credit program. Thus, government remains a major partner in these efforts (see Department of City Planning, 1995).

The city's roughly 200 nonprofit community development corporations (CDCs) are responsible for developing and maintaining about 100,000 housing units, two and one-half times as many units as are owned by the city and more than half the total number of public housing units. CDCs are only one part of a large sector of community-based organizations that are playing an increasingly important role as a pillar of civil society supposedly standing between public and private sectors, and between government and individual households. Scores of local economic development corporations sponsor programs for small business development and assistance, and there are hundreds of community-based organizations and block associations dedicated to everything from pocket park development to crime prevention.

Despite the significant role they play in neighborhood development, the public and nonprofit housing sectors continue to operate at the margins. Public and nonprofit development has helped individual households, buildings, and blocks without necessarily helping neighborhoods. Limited resources are available in these sectors, and public support has been heavily skewed toward physical development—filling up vacant lots and buildings—and not necessarily preserving communities. Coordination of city agencies with neighborhoods is uneven, and the main formal mechanism for coordination—the 59 appointed community boards—is underfunded and understaffed. Each board covers an average area of 100,000 people with a staff of no more the two or three full-time persons, and some community boards include half a dozen clearly identifiable neighborhoods.

Even though there has been significant development in distressed neighborhoods in recent decades, the majority of resources for community development in the city are in private sector projects, particularly the downtown mega-projects like the World Financial Center, Times Square Revitalization, Donald Trump's Riverside South, and Queens West. Private developers, who continue to invest the vast majority of their collective wealth in wealthier residential enclaves and business districts, command the largest sums in circulation and move into distressed neighborhoods only when there are subsidies involved or poor people have left. The much-heralded "marketplace" is virtually absent in the most distressed neighborhoods.

## THE MOST DISTRESSED NEIGHBORHOODS:
## THE CASE OF RED HOOK

In general, government-sponsored rehabilitation and new housing de-velopment since the 1970s have helped stabilize and improve some low-income neighborhoods and parts of neighborhoods. Some of the most distressed neighborhoods in New York City, however, were by-passed by development. Some of these neighborhoods were located far from the path of urban renewal but faced a sort of reverse gentrification. Others are physically isolated and consequently have static or declining property values and little potential for future private investment. Most are communities of color.

Some of these neighborhoods missed out on residential development opportunities because of the presence of industry and a relatively limited inventory of vacant city-owned land and buildings. Red Hook is one such neighborhood. It is a relatively small, geographically iso-lated community that was devastated by the first phase of government intervention in the 1950s and 1960s (Act One) and completely bypassed by development in the second phase (Act Two).[1]

Red Hook is a small, segregated, and isolated community of 11,000 people on the South Brooklyn waterfront. Seventy percent of its popu-lation lives in Red Hook Houses, a public housing project of mostly mid-rise buildings. About half of the population is African American, and 42% is Latino, mostly Puerto Rican. The community is physically divided between the public projects and the area known as "The Back," where a mixed ethnic population lives in small rental buildings and homes. The residential community is surrounded by a small but diverse industrial community, including warehousing and manufacturing. Wa-terfront land includes industrial and maritime activities, and there is limited public access to the waterfront even though the community is surrounded by water on three sides.

Until the 1950s, Red Hook was a thriving industrial, maritime, and residential community whose livelihood depended on robust port ac-tivity. Containerization revolutionized the shipping industry, but politi-cal decisions made by the Port Authority of New York and New Jersey led to the siting of all but one container facility on the New Jersey side of the New York Harbor. The container facility in New York was a small one serving niche markets, and it was located on a small portion of Red Hook's waterfront. The threat of condemnation that would have led to a larger port facility, combined with the decline of many maritime-related industries that hired local residents, contributed to residential abandonment in Red Hook. Between 1950 and 1990, Red Hook's population declined by almost half.

Robert Moses–era public works projects financed with federal assistance contributed to Red Hook's transformation in the immediate postwar era. One of the first public housing projects in the city was built on Red Hook land once occupied by homeowners and squatters. At about the same time, an expressway was built, cutting off the projects and declining industrial area from upland neighborhoods that were part of Red Hook—they were soon to be dubbed Carroll Gardens and Columbia Street.

Growth spurred in Act Two of the urban policy drama eluded Red Hook because opportunities for public or private investment were limited by land speculation. Because of its ready access to the expressway, downtown Manhattan, and Brooklyn, vacant land and buildings in Red Hook were snapped up quickly at auction by small speculators and industries, including truckers, waste haulers, and auto storage dealers. As a result, after the first wave of abandonment, there were very few city-owned vacant lots and buildings in Red Hook. Without vacant land and buildings, there was little room for the city's housing development programs. By 1990, Red Hook had one of the largest concentrations of waste transfer stations in the city, a distinction that made the community known as a prime example of environmental racism.

Without much vacant city-owned property today, the potential for subsidized redevelopment in Red Hook remains limited. Because of its real and perceived isolation, land values remain low enough to preclude private residential investment. Seventy percent of the population lives in public housing, where until recently there has been little need to worry about displacement and little interest in new residential development.

Even in severe distress, optimism is possible. In some ways, people in Red Hook have started to fulfill Jesse Jackson's exhortation to "keep hope alive." The completion of a Red Hook *Plan for Community Regeneration* in 1994 was part of a process of increasing community activism and optimism about the future. The plan seeks to break down community divisions by encouraging new housing development for a diverse community, public access to the waterfront, preservation of a mixture of industrial and residential uses, and physical integration of Red Hook Houses with The Back. The plan was written after Red Hook groups including tenants, homeowners, and businesspeople defeated a plan by the city to build two large sludge treatment plants on the waterfront, and after the death of Patrick Daly, revered principal of one of Red Hook's elementary schools, who was caught in a cross fire of drug dealers. This brought people together in mourning and protest. Thus, optimism grew out of outrage, struggle, and organizing.

Historical patterns of racial discrimination within city government reinforce the powerlessness of the most isolated minority neighborhoods. Racial divisions within Red Hook also contribute to its powerlessness, but in recent years tenants in Red Hook Houses have begun to develop partnerships with homeowners in The Back. Several hundred households in Red Hook Houses have incomes high enough to afford home ownership, and they see future residential growth as of possible benefit to themselves.

In today's political climate of privatization and deregulation, it would be unrealistic to expect much support from government for community development anywhere, much less in Red Hook. The good news is that the massive public works and displacement that created today's Red Hook are universally vilified and probably a thing of the past. The bad news is that government policies and programs that assisted other neighborhoods and businesses will not be available to help the neglected, severely distressed communities like Red Hook that were bypassed in recent decades. Under the current Republican mayor, the city has said it will stop acquiring abandoned lots and residential properties—even though several thousand housing units are abandoned every year, not as high a rate as in the 1970s, but higher than in the 1980s. The housing agency is also phasing out programs to rehabilitate abandoned buildings for low-income tenants.

Government assistance to neighborhoods has not ended, but it is increasingly being targeted to "economic development," or, one might say, to private business development. Although the city and state have awarded tax reductions to waterfront industries in Red Hook, there are few programs for preservation and development of housing.

Looking toward the future, the next wave of government policies could very well devastate any hopes for integrated development and an improved quality of life in Red Hook. Privatization of public housing, should the New York City Housing Authority follow Washington's lead, would likely result in the displacement of low-income households without offering them alternative opportunities within the community. Even a partial privatization, without government commitment that there be no net loss of low-income units, might achieve integration within projects, but at the expense of equity and integration within the neighborhood. Federal and local cuts in public assistance, education, and health care will be especially felt in Red Hook because of the large proportion of the population that depends on this assistance.

On the immediate horizon, Red Hook residents and businesses are concerned about the city's recent announcements that it is seeking to revive the maritime industry on the Brooklyn waterfront, including Red Hook. This would include construction of a rail freight tunnel from a point just south of Red Hook to Staten Island or New Jersey, and reclamation of waterfront land for shipping. It is not clear whether the city will go back to the old, discredited policies of clearance to achieve its objectives, continue the present policies of neglect, or try a new approach involving cooperation and integration of maritime and residential areas.

Thus, like many severely distressed neighborhoods in New York City, Red Hook is facing an uncertain future.

## ACT THREE: THE UNCERTAIN FUTURE

As federal and local government initiatives change and are cut back in the 1990s, the most vulnerable neighborhoods are being left on their own to protect themselves. Controls on rents and real estate development that have cushioned the processes of gentrification and displacement are being eased. Optimism about the future of poor neighborhoods must be tempered by the fear that the social safety net that many need to survive will be completely shredded in coming years.

Federal contributions that found their way to distressed neighborhoods via welfare, Medicaid, and food stamps are dwindling rapidly. About one-third of New York City's population consists of immigrants, many of whom will be affected directly by the 1996 welfare "reform" bill. The curtailment of public assistance to immigrants will significantly reduce the money in circulation in many low-income neighborhoods, further increasing dependence on the underground earnings from drugs and crime. Although funds for housing and community development have not declined drastically, there is no expectation that current levels of financial support from Washington will withstand the continuing efforts at downsizing related to the bipartisan effort to balance the federal budget.

Under the second Clinton administration and Republican Congress, the future of spending in most of the city's successful housing programs is uncertain. As noted in the city's *Proposed Consolidated Plan* (Department of City Planning, 1995, p. I-1), "it is likely that HUD funding will be cut. . . . It is federal assistance which makes it possible for joint ventures . . . to expand and improve New York City's affordable housing stock."

Anticipating declining revenues from Washington, and under a mayor with a penchant for privatization, the city's Department of Housing Preservation and Development has intensified its perennial attempts to sell off the remaining city-owned residential buildings. Without recourse to the city to pay for repairs and maintenance, low-income tenants will either suffer worse conditions, pay higher rents, or be forced out. In the private housing market, low-income tenants are constantly threatened by attempts by the state legislature to eliminate all rent regulations. Seventy percent of New York City's households are tenants. Although almost half are paying more than 30% of their incomes for rent (Blackburn, 1993, p. 283), without rent regulations, many more families would be forced to double up or join the homeless population on the streets.

In recent years, there have been significant losses in housing stock because of abandonment by private landlords—more than half the rate of loss during the worst years in the 1970s. Between 1991 and 1993, 19,000 housing units were abandoned (Blackburn, 1993, p. 135). At the same time, the city has stopped acquiring tax delinquent properties, foreclosing any future publicly sponsored renovation of abandoned buildings. The myth that the private market can and will pick up the slack is just that—a myth.

## THE EMPOWERMENT ZONE

The main urban program developed under the Clinton administration—Empowerment Zones (EZ)—has yet to have any impact in New York City because it just went into operation. The New York City EZ, which includes portions of Harlem and the South Bronx, has taken longer to get off the ground than other EZs because of bickering between the Republican-controlled city and state governments, neither of which have demonstrated much interest in seeing the program move forward in heavily Democratic neighborhoods, especially Harlem, the home of Representative Charles Rangel, who authored the EZ legislation.

It is difficult to imagine how the EZ's tax benefits to employers, and the array of services it will support, will seriously influence poverty in the target neighborhoods. The tax benefits are likely to favor large retailers like the Disney Corporation, which already made plans to move to 125th Street in Harlem before EZ came into being. This new rush of investment may help fulfill the dream of Harlem's business elite to make the area a center for international trade and commerce, but the biggest beneficiaries are most likely to be corporations from outside Harlem and not locally grown capitalists. The city's Republican mayor, who is not a popular person in Harlem, is already preparing his EZ team to shower praise on him for spurring Harlem's renaissance by standing up to the community's Democratic elite and blaming them for past failures. This thinly veiled paternalism reinforces the institution of racism that historically has contributed to Harlem's enduring problem as a second-class neighborhood in a global city.

EZ tax benefits will not create development. They are likely to subsidize retailers who are taking income out of the community instead of helping to generate income within the community. They are not likely to attract jobs for skilled labor. In fact, new jobs in the retail and service sectors are often located in former industrial areas, where better-paying manufacturing jobs are being forced out by increasing property values.

The service programs to be supported by the EZ may well contribute to a gradual improvement in the quality of life for EZ residents and help buttress struggling community-based organizations. While directly benefiting some community members, however, they cannot compensate for the serious local government cutbacks in school expenditures, health services, and housing maintenance. In sum, the EZ appears to be a continuation of the string of ambiguous policies that began in the 1970s and that may very well be the final act in the urban policy drama.

## NOTE

1. This analysis is based on the author's more than 5 years of work in Red Hook as a planner. The author provided assistance in the preparation of a plan for the neighborhood (Community Board 6, 1994) and continues to work with community-based organizations there.

# CHAPTER 12

# FUTURE PROSPECTS FOR DISTRESSED URBAN NEIGHBORHOODS

W. Dennis Keating
Norman Krumholz

## INTRODUCTION

With their economy and social fabric frayed by poverty, racial segregation, and other vast social forces, the older American industrial cities
have been in sharp decline. Almost all are losing population and economic investment and are becoming homes to the poorest and most
dependent residents of the metropolitan areas. Much of the population
remaining in these cities lives in neighborhoods that provide poor
educations, have high crime rates, and in every respect offer a lower
quality of life than that enjoyed by other Americans.

Local and national leaders typically has responded to this crisis by
attempts to stimulate investment, bring middle-class people back to the
city, and develop heavily subsidized real estate projects in downtown
areas, hoping that the benefits of new growth will "trickle down" to

those in the lower reaches. To an extent, their efforts have succeeded as new office buildings, hotels, and sports stadiums have reshaped the skylines in urban America.

## WHAT HAVE WE LEARNED?

The strategy has not succeeded in reducing poverty, unemployment, or dependency among the cities' resident populations, and it has left untouched the widening economic disparities between central cities and their suburbs, where median family incomes and rates of job growth and employment typically are much higher than in the central cities. The "trickle down" strategy has produced few, if any, benefits for increasingly destitute city residents. Instead, it has produced more bifurcated cities and regions, sharply split between downtown and other zones of relative affluence and low-income and working-class city neighborhoods.

Given the scale of these seemingly intractable problems, these older industrial cities were said to be on the way to permanent obsolescence. Instead of sliding into the dustbin of history, however, some cities and some neighborhoods stood and fought. Although their problems remain substantial, their successes as reported in this book and other sources (e.g., Vidal, 1992, 1996) provide evidence that it is possible to rebuild, even in the face of daunting odds.

Most of the examples in this book are cities and neighborhoods that are predominantly black: Atlanta's Peoplestown; Camden, New Jersey; East St. Louis, Illinois; Chicago's North Lawndale, Cleveland's Hough and Central, most of Detroit's Empowerment Zone, and Miami's Overtown. The examples of South-Central in Los Angeles and Red Hook in New York City are racially and ethnically mixed.

All these examples face overwhelming and familiar problems related to poverty. All have a past history of governmental intervention, mostly through federal programs. Unfortunately, these programs have not solved their problems. Faced with the prospect of declining federal urban aid to central cities and their poorest neighborhoods, these areas and their counterparts in other U.S. cities must seek substitute sources of assistance.

Certainly, the community development corporation (CDC) movement is an example of the resilience of poor neighborhoods and the critical importance of community involvement in any revitalization efforts. As the examples of Atlanta's Peoplestown Revitalization Corporation, Chicago's Lawndale Christian Development Corporation, and the Winstanley/Industry Park Neighborhood Organization in East St. Louis show, CDCs are instrumental in both defending and redeveloping distressed urban neighborhoods. Any urban revitalization strategy, nationally and locally, must support community-based initiatives like CDC-planned programs.

CDCs alone, however, cannot rebuild these central city neighborhoods. Past CDC failures in Watts (Los Angeles) and Hough (Cleveland) attest to this reality. Whatever federal or local programs are available to assist neighborhoods, their leaders must be involved. Several of the examples in this book show that neighborhood leaders have not been involved effectively in the U.S. Department of Housing and Urban Development's (HUD) Empowerment Zone/Enterprise Communities (EZ/EC) program, either in developing the city's proposal or its early implementation. This was notable in cities like Camden, Los Angeles, and New York. The neighborhood movement has had only very limited success in gaining political influence at the municipal level.

## PROGRESSIVE MUNICIPAL GOVERNMENT

There are examples of municipal governments under progressive leadership seeking to address the problems of their poorest citizens and the neighborhoods in which they reside. Two notable examples are Chicago under the late Mayor Harold Washington (Clavel & Wiewel, 1991; Giloth, 1996; Mier, 1993) and Boston under Mayor Ray Flynn (Dreier, 1996). In both cases, community organizations both supported policies to improve the lives of poor and disadvantaged residents and also pressured the city government to initiate and implement reforms that challenged the interests of entrenched pro-growth business interests and their governmental allies. Ferman (1996) analyzes this in Chicago, focusing on such policies as the plan called Chicago Works Together, a centerpiece of Mayor Washington's proposed economic development reforms. Robert Giloth, who worked for Washington, recounts the impact of such policies in Chicago's North Lawndale neighborhood.

In Boston, a noteworthy example is the Dudley Street Initiative (DSI). This is a community-based effort in Roxbury, one of Boston's most distressed neighborhoods, that began in 1984-1985 to try to address comprehensively all the many problems confronting this predominantly minority neighborhood. As a result of redlining, disinvestment, and poverty, the neighborhood faced substandard and abandoned housing, vacant lots and inadequate services, and unemployment. DSI pressured the Flynn administration to assist it in overcoming these problems. The most dramatic example of this alliance was Flynn's support for a unique redevelopment policy. Under pressure, the Boston Redevelopment Authority authorized DSI to use the power of eminent domain, if necessary, to reclaim abandoned properties for redevelopment in accordance with a comprehensive community-based plan (Medoff & Sklar, 1994).

## FEDERAL DEVOLUTION

City government is likely to play an increasingly important role in the restoration of distressed urban neighborhoods. The continuation of cutbacks in federal urban aid and the devolution of many key federal social programs to state and local governments point in that direction. Cities are likely to have reduced resources to meet these challenges. Also, as Rich (1993) showed in his analysis of the impact of HUD's Community Development Block Grant (CDBG) program, local governments are less likely than the federal government to target aid to the poorest people and neighborhoods. Absent federal social targeting priorities that mandate this, the politics of local governments with broad discretion are likely to result in the dispersion of such funding widely throughout city neighborhoods and downtown. The EZ/EC program may be one of the few federal initiatives of the 1990s to mandate such targeting to neighborhoods of concentrated poverty. As of late 1997, the final evaluation of the first phase of this program had not yet been released. The early EZ experience is discussed in the chapters on Camden and Detroit. The authors of other case studies (e.g., Cleveland, Los Angeles, and New York) speculate about its likely future impact.

## THE URBAN POLICY DEBATE

Two prevailing political philosophies color the policy discussions of the decline of severely distressed urban neighborhoods and their poor inhabitants. Many persons on the left of the political spectrum believe that past federal programs have failed to be effective because of a lack of resources and political will. All that is needed, they argue, is a more resolute commitment to adequate education, job training and development, housing, health care, and drug prevention and treatment. To repeat a commonplace belief from the 1960s, the war on poverty has not been won because it was never fought with sufficient compassion and resources to make a difference.

Conservatives, on the other hand, take the opposite position, that regardless of the level of spending, little is possible; that federal resources are ineffective when pitted against the inner city's crime-ridden mean streets, indifference to the classroom, and decay of the family and social mores. The best policy, they argue, is no policy at all, certainly no policy that would lavish public money on destructive behavior that nobody understands or knows how to change.

Neither of these positions is likely to dominate urban programs or public policy as America moves toward the year 2000. Yet, if the 1996 presidential campaign is any indication, the tilt is heavily toward diminished federal governmental intervention. President Bill Clinton's reluctance to make any explicit statements about the condition of poor people in poor neighborhoods, the budget-balancing fervor of both the Clinton administration and the Republican Congress, the deliberate "reinvention" or downsizing of HUD announced in 1995, and the overwhelming emphasis on the needs of the middle class all suggest the probability that major new federal commitments to resolve the problems of poor people in distressed urban neighborhoods will not be forthcoming.

Instead, urban interventions in the near future are likely to resemble President Clinton's Empowerment Zone program, which was based in part on President Reagan's Enterprise Zone proposal championed by Jack Kemp, secretary of HUD under President Bush. Both depend heavily on the notion that taxes and regulations are inhibiting economic growth in distressed neighborhoods. Both programs rely on tax credits, most of them targeted at employers to unleash the transforming power of the free market and generate economic development, jobs, and income for neighborhood residents and businesses. Both programs emphasize tax reductions and the elimination of governmental regulations in the hope that these measures will solve, or at least ameliorate, the worst economic problems of blighted communities, but the Clinton program also addresses social problems. Time will reveal whether this approach will be successful in coping with persistent and concentrated poverty, poor health, and inadequate education, and the litany of human misery found in distressed neighborhoods, or whether real reform will require a much more significant intervention by the federal government than most Americans seem to prefer.

Harvard University economist Michael Porter offers another dimension of the policy debate, arguing for a market-based development strategy in inner cities (Porter, 1995). Porter points to job opportunities for poor residents. He criticizes past governmental efforts to revitalize inner-city neighborhoods while advocating governmental assistance to stimulate private investment. His critics have sharply attacked his position, including his assumptions about the types of job opportunities possible and his views about the efficacy of governmental policies (Boston & Ross, 1997).

Another aspect of the debate, one that has persisted, is whether public and private investment in downtown redevelopment has neglected poor urban neighborhoods or whether it has both revitalized many central cities and provided employment, both to suburban white-collar commuters and city residents. Case studies that address this controversy abound (e.g., Fainstein, Fainstein, Hill, Judd, & Smith, 1983; Ferman, 1996; Squires, 1989). This issue is featured in several of the case studies in this book as well (Atlanta, Cleveland, and Miami).

## PEOPLE VERSUS PLACES REVISITED

The people versus places debate also persists. Whether to try to help the poor in their central city neighborhoods or whether to assist them in improving their lives so as to enable them to leave concentrations of poverty and move to mainstream neighborhoods, both in the inner city and in the surrounding suburbs, remains a much-debated topic. The

1968 Kerner Commission report advocated both approaches, as have others since. Particularly active in this discussion and urging metropolitan cooperation and regional mobility for the urban poor are David Rusk (1993), Anthony Downs (1994, 1997), and Paul Jargowsky (1997). Some see little advantage of one strategy versus the other (Boger & Wegner, 1996).

A related debate has been characterized as race versus class. Sociologist William Julius Wilson (1996) has argued in favor of universal approaches to social problems such as long-term unemployment. In recognition of the political backlash against targeted place-based programs and affirmative action policies favoring racial minorities, Wilson believes that the only hope for these groups is to share in general social policies that will also benefit them, even if they do not receive priority or preference. At the moment, the political tide is running against these types of universal social entitlements, as well as affirmative action policies.

A contrary view is that of sociologists Douglas Massey and Nancy Denton (1993). They argue that the key feature of urban poverty is de facto racial segregation that traps so many minorities in inner-city neighborhoods. Rather than the promotion of programs and policies to alleviate poverty, they argue for a range of expanded policies to fight racial discrimination, particularly in housing. Without this, they conclude that poor minorities will remain trapped in inner-city ghettos. Halpern (1995) and the 1990 Committee on National Urban Policy have echoed that view. They and others demonstrate that widespread racial discrimination and segregation in housing is persistent, but there are few signs of support for an active federal policy to attack these problems, especially the exclusionary land use policies that characterize most predominantly white suburban communities. Given the absence of a constituency for either approach, it is hardly surprising that the most promising efforts have focused on the rebuilding of urban neighborhoods, whether racially segregated or not.

Race is the critical issue in this debate. Sociologists Douglas Massey and Nancy Denton (1993) argue that pervasive racial segregation is the most important factor in the continuation of racial ghettos characterized by concentrated poverty. They argue for accelerated efforts to attack racial discrimination in housing and expand the opportunity for regional mobility for minorities. While middle-class minorities have been able to leave central city neighborhoods for the suburbs (although not necessarily to racially diverse suburban areas), poor minorities have not had the income necessary for such a move in the absence of lower-priced rental housing in the suburbs. Short of remedial litigation against them, predominantly white suburban governments usually have not been willing to promote racial diversity. The exceptions to this prevailing pattern are few (Keating, 1994). Although expanded enforcement of fair housing legislation would somewhat increase the minority suburban population, it would not by itself lead to greatly increased mobility for the residents of concentrated poverty areas.

Likewise, it is unlikely that many whites would consider moving to poor central city neighborhoods, absent tremendous improvement in conditions. The Harborpoint mixed-income housing project in Boston, formerly an isolated and troubled public housing project, is a rare example of this. Instead, inner-city racial diversity still occurs, usually during a transition from white to black and minority in a segregated housing market. The examples of long-term racially diverse central city neighborhoods are few (Nyden, Maly, & Lukehart, 1996; Saltman, 1990). The examples of Atlanta's Peoplestown, Chicago's North Lawndale, and Cleveland's Hough in this book are but three examples of racial transition, followed by resegregation. When middle- and upper-income newcomers are attracted to areas occupied by the poor, including minorities, the experience usually has been gentrification, with the result that eventually few, if any, of the former residents remain as rising housing prices and land values force them out. As Norman Krumholz notes in Chapter 6, there has been a small scale influx of middle-income black home buyers into Cleveland's impoverished Hough neighborhood, although so far there are no signs of gentrification.

## REBUILDING DISTRESSED URBAN NEIGHBORHOODS

### Federal Policy

Under these circumstances, what can be done in the future to rebuild these neighborhoods? The least likely scenario would be a repetition of the 1960s. Barring massive social unrest, a downsized federal government that has promised to balance the budget, while also reducing taxes, is not going to restart a wide range of programs targeted at the inner-city poor and their neighborhoods. Instead, reduced federal aid increasingly will be channeled through state and local governments, which will have considerable discretion in its distribution. Federal policies may be, on balance, more destructive than constructive, depending on the outcome of the new initiatives identified above.

Despite President Clinton's naming of a federal advisory board on race relations in June, 1997, the federal government has shown no signs of initiating policies designed to promote further racial integration in housing, schools, and neighborhoods. With political and legal attacks on such policies as school busing for desegregation and affirmative action in employment and education, along with black majority congressional voting districts gaining momentum, there has not been a national public dialogue about the continued existence of highly segregated urban neighborhoods where there is a concentration of the poor or an "underclass."

## Community-Based Revitalization

If there is little likelihood of expanded federal urban aid or a reduction in racial segregation coupled with concentrated poverty, then what are the prospects for these neighborhoods? As the case studies show, where there are strong community-based organizations, there is hope for the betterment of the neighborhood. Using a combination of public and private funds, progress has been made, in rebuilding both the social fabric and the physical features of these neighborhoods. Even so, there are limits as to how far this can go in addressing needs. The ambitious goal of the Sandtown-Winchester project—to transform this impoverished Baltimore neighborhood within a decade—is unusual. The increasing number of community building initiatives gives hope that this approach will grow in acceptance.

A key element in attaining marked improvement in these neighborhoods is the capacity of the CDCs. Stoecker (1997) argues that CDCs have a limited financial and organizational capacity to effectively address the needs of those neighborhoods in which they operate. He argues for much enlarged CDCs and separate organizations to organize and empower poor neighborhood residents. Bratt (1997) and Keating (1997) have taken issue with this analysis; however, it must be conceded that CDCs with very limited resources cannot be expected to plan and then implement the comprehensive redevelopment of their neighborhoods. There must be major involvement by government, as well as private intermediaries (Keyes, Schwartz, Vidal, & Bratt, 1996).

What could emerge is a more modest version of what Downs and others advocate on a comprehensive scale. In a select number of neighborhoods, well-organized networks of community organizations will develop comprehensive neighborhood plans. They will then utilize a combination of governmental social service and redevelopment programs at all levels, corporate contributions, and philanthropic grants to make improvements. This can improve the lives of many in the neighborhood, as CDCs have demonstrated over the past few decades. Effective neighborhood leadership will be a key element of this process. The tactics of these organizations may be collaborative or confrontational.

## Creating Jobs

What CDCs, other community organizations, churches, and social service agencies have not done and cannot do is create jobs on the scale that is needed in these communities. Although they can sponsor and participate in job training, adult education, and related programs, employment must come from the private sector or, as a last resort, government. At the federal level, with the economy stable and unemployment rates nationally and regionally at low levels without accompanying inflation, there is no support for a publicly funded jobs program, even for former welfare recipients required to find work. To compound this problem, although the economy generated millions of new private sector jobs in the 1990s, all too many were low-paying, without benefits, and part-time or seasonal (Nightingale & Haveman, 1995). In an increasingly service-oriented economy, there is no longer the possibility of a large number of low-skill, industrial jobs being generated for the unemployed residents of urban areas of concentrated poverty. The case study cities all have this problem (although Miami never had an industrial job base).

Where will the jobs be found? Without work, as Wilson (1996) emphasizes, there is little hope that other major social ills can be successfully addressed. Kasarda and Ting (1996, p. 414), after analyzing urban joblessness and poverty in 67 large U.S. cities, point to 10 policy prescriptions to reduce structural barriers and to improve mobility and job access by the inner-city advantaged.

Among these are educational reforms. To obtain work in the legitimate economy, the poor must be literate and at least educated enough to obtain a high school degree. This inevitably leads to the serious problems of public school systems, beset by funding problems, huge dropout rates, deteriorating buildings, demands for racial balance, and the threat of private competition through voucher subsidies. In Baltimore, a private company took over management of public schools; in Chicago, it was the mayor; in Cleveland, it was the state of Ohio through federal court order. This indicates the crises that have overwhelmed much of the public school system in central cities. How the schools serving the poorest children living in the poorest urban neighborhoods can be improved remains a formidable and unsolved question.

Welfare reform legislation, passed into law in 1996, is a major unknown. Within the following 5 years, millions of welfare recipients will lose their benefits and must try to find work. This particularly affects female-headed households with children. If many are unable to find employment or if the employment they find does not pay an income adequate for shelter, day care, transportation, and the other necessities of life, the consequences will be devastating. To compound this threat, many of these same households face cutbacks or elimination of other existing subsidies, including those for housing, food stamps, and medical care. Local governments will be hard pressed to fulfill the requirements of federal and state welfare reform laws to assist these households. Should these households lose all assistance, it is likely to be local governments, charitable agencies, churches, and community-based organizations that will have to try to assist them, especially the children. In the EZ/EC neighborhoods, the federal cutbacks may be so overwhelming that they may swamp any positive effects of the job training and development efforts now under way.

Finally, crime and drugs confound the revitalization of the poorest neighborhoods. If crime and drug use abound, whether as a sign of despair or simply as a solution to survival, then efforts to rebuild the physical and social infrastructure of poor neighborhoods are likely to fail, as those who can move to safer locations will do so. Despite increased funding for community policing and falling crime rates in U.S. cities in 1997, crime remains a major impediment to the rebuilding of neighborhoods. Community policing and neighborhood crime watches can have an effect in reducing crime. Both require the involvement of neighborhood residents.

## THE FUTURE OUTLOOK

If public education can be improved, jobs can be found, and urban neighborhoods can be made safer, then there is reason to believe that poor communities can be rebuilt. This is a tall order indeed. Likewise, if the focus is also to increase the mobility of the residents of the poorest urban neighborhoods to better areas, the required steps also will be difficult. Necessary steps, such as enforcing fair housing and employment laws and providing improved transit to open job opportunities, would require substantial changes in suburbia.

Despite the complexity of these challenges, the alternative is to sanction the continued existence of highly segregated and isolated neighborhoods with high poverty concentrations, in which hopelessness and social and physical disintegration prevail. This accepts the dual city of the prosperous and the poor. It consigns large numbers of city dwellers to neighborhoods characterized by all the signs of decline and, ultimately, abandonment—vacant rubble-strewn lots and empty, crumbling, graffiti-ridden buildings, both residential and commercial; poor or few public services and facilities; high crime rates; and substandard and overcrowded housing and schools. Because we have the resources and the institutions to correct these conditions, it should be unacceptable that we tolerate their continued existence. To find the political will to eradicate these conditions of human distress remains a great challenge.

The optimal approach would be the kind of sweeping reforms and universal approaches advocated, for example, by Goldsmith and Blakely (1992). They point to such policies as universal health care, family income support for the working poor, merging welfare with Social Security, adopting a national industrial policy, and a national emergency employment program. Wilson (1987) argued for such universal entitlement programs as child support and family allowances and greatly expanded national welfare standards. In the wake of the 1994 and 1996 national elections, however, political support for expanding federal support for the poor and distressed urban areas has all but disappeared. This reflects the lack of political power possessed by either the poor or the cities where they mostly live. In addition, in the era of proposals for a balanced federal budget and further tax cuts, entitlements are being cut back or eliminated, not expanded.

We can only hope in the near future for the survival of remaining federal programs and policies that support the revitalization of poor urban neighborhoods and their residents. We can hope that more progressive local governments can forge public-private partnerships that address these problems, rather than focusing primarily on the development of central business districts to the detriment of distressed urban neighborhoods.

The real job of rebuilding urban neighborhoods, however, must begin at the grass roots. This means getting more people involved in the gritty, grimy job of politics. It means convincing good people to run for office and creating strong alliances among the poor, the near-poor, unionized and non-unionized workers, community-based groups, universities, and others. It means committing time and effort to support of initiatives that stress values and not just programs. In short, it means action at the grass roots if fundamental changes are to occur.

# REFERENCES

Ambrose, Brent W., Hughes, William T., Jr., & Simmons, Patrick (1995). Policy issues concerning racial and ethnic differences in home loan rejectiln rates. *Journal of Housing Research, 6*(1), 115-135.

Anderson, A., & Pickering, G. W. (1986). *Confronting the color line: The broken promise of the Civil Rights Movement in Chicago.* Chicago: University of Chicago Press.

Anderson, Martin. (1964). *The federal bulldozer.* Cambridge, MA: MIT Press.

Angeles, Mark. (1994, October 4). Camden to keep 10 million grant. *Courier Post,* pp. A1-A5.

Argyris, Chris, Putnam, Robert, & McLain Smith, Diane. (1987). *Action science.* San Francisco: Jossey-Bass.

Baldassare, Mar. (1994). *The Los Angeles riots: Lessons learned for the urban future.* Boulder, CO: Westview.

Baron, Harold M. (1974). Building a Black community: Popular economics in Lawndale. *FOCUS/Midwest, 11*(69), 18-22.

Beck, S. A. (1987). The limits of presidential activism: Lyndon Johnson and the implementation of the Community Action Program. *Presidential Studies Quarterly, 17*(3), 541-566.

Bellamy, John Stark, II. (1997). *The maniac in the bushes: And more tales of Cleveland woes.* Cleveland, OH: Gray.

Bellush, Jewel, & Hausknecht, Murray (Eds.). (1967). *Urban renewal: People, politics and planning.* New York: Anchor.

Blackburn, Anthony J. (1993). *Housing New York City.* Department of Housing Preservation and Development, City of New York.

Boger, John Charles, & Wegner, Judith Welch (Eds.). (1996). *Race, poverty and American cities.* Chapel Hill: University of North Carolina Press.

Borenstein, Seth. (1994, February 17). Segregation posts decline, but still high. *Miami Herald*, pp. 1A, 24A.

Boston, Thomas D., & Ross, Catherine L. (Eds.). (1997). *The inner city: Urban poverty and economic development in the next century*. New Brunswick, NJ: Transaction Publishing.

Bowden, C., & Kreinberg, L. (1981). *Street signs Chicago: Neighborhood and other illusions of big-city life*. Chicago: Chicago Review Press.

Bowley, Devereux, Jr. (1978). *The poorhouse: Subsidized housing in Chicago, 1895-1976*. Carbondale: Southern Illinois University Press.

Bratt, Rachel G. (1997). CDCs: Contributions outweigh contradictions, a reply to Randy Stoecker. *Journal of Urban Affairs, 19*(1), 23-28.

Bronstein, B. (1996, March 12). Management group buys minority bank. *American Banker*, p. 6.

Brown, P., Pickens, L. M., & Mollard, W. (1996a). *Report on the North Lawndale initiative: The Steans Family Foundation*. Chicago: The Chapin Hall Center for Children at the University of Chicago.

Brown, P., Pickens, L. M., & Mollard, W. (1996b). *The Steans Family Foundation: Work to date*. Chicago: The Chapin Hall Center for Children at the University of Chicago.

Burgess, E. W. (1925). The growth of the city. In R. E. Park, E. W. Burgess, & R. D. McKenzie (Eds.), *The city* (pp. 47-62) Chicago: University of Chicago Press.

Burton, C. Emory. (1992). *The poverty debate: Politics and the poor in America*. Westport, CT: Greenwood.

Caraley, Demitrios. (1992). Washington abandons the cities. *Political Science Quarterly, 107*(Spring), 1-30.

Carmon, N. (Ed.). (1990). *Neighborhood policy and programmes: Past and present*. New York: St. Martin's.

Casas, R., & Colvin, T. (1974). *Lawndale: How far is up*. Evanston, IL: Medill School of Journalism, Northwestern University.

Castro, Max J. (1992). The politics of language in Miami. In Guillermo J. Grenier & Alex Stepick III (Eds.), *Miami now! Immigration, ethnicity and social change* (pp. 109-132). Gainesville: University Press of Florida.

Catlin, Robert A. (1993). *Racial politics and urban planning: Gary, Indiana, 1980-1989*. Lexington: University of Kentucky Press.

Challenges of 1997. (1997, January 1). *Detroit News*. Retrieved from the World Wide Web: http://www.detnews.com/EDITPAGE/9701/01/1edit/1edit.htm

Chicago Fact Book Consortium. (1995). *Local community fact book Chicago metropolitan area 1990*. Chicago: The Chicago Fact Book Consortium, The University of Illinois at Chicago.

Chicago Rehab Network. (1993). *The Chicago affordable housing fact book: Visions for Change*. Chicago: Chicago Rehab Network.

*Chicago Tribune*. (1986). *The American millstone: An examination of the nation's permanent underclass*. Chicago: Contemporary Books.

City of Detroit. (1994). *Jumpstarting the Motor City: With new ideas, new relationships and new technologies for the Detroit Empowerment Zone* (Vol. 1). Detroit: City Planning Department.

City of Miami. (1996, July 23). *Special election, cumulative report—final*. Miami: Author.

City of Miami Planning, Building and Zoning Department. (1993). 1975-1993 projects: Overtown Community Development Target Area. In *Neighborhood Planning Program, 1994-96* (pp. 45-49). Miami: Author.

City of Miami Police Department. (1996, April 11). *Part I UCR crimes by MPD net area by dispatch date* (Reports for calendar years 1994 and 1995). Miami: Author.

*Civic Vision 2000 Citywide Plan.* (1991). Cleveland: City Planning Commission.

Clarke, Susan E. (1984). Neighborhood policy options: The Reagan agenda. *American Planning Association Journal, 50*(4), 493-501.

Clavel, Pierre, & Wiewel, Wim (Eds.). (1991). *Harold Washington and the neighborhoods: Progressive city government in Chicago.* New Brunswick, NJ: Rutgers University Press.

Cleveland Council of Economic Opportunities. (1997). *Total dollar value of welfare cuts, food stamps and general assistance by Cuyahoga County SPAS.* Cleveland: Author.

Community Board 6. (1994). *Red Hook: A plan for community regeneration.* New York: Author.

Connell, J. P., Kubisch, A. C., Schorr, L. B., & Weiss, C. H. (1995). *New approaches to evaluating community initiatives.* Washington, DC: The Aspen Institute.

Conot, R. (1974). *American odyssey.* New York: William Morrow & Company.

Costello, N. (1994, December 12). Teamwork is key. *Detroit Free Press,* p. A10.

Costigan, Patrick. (in press). *Building community solutions.*

Cummings, Scott C., Koebel, Theodore, & Whitt, J. Allen. (1989). Redevelopment in downtown Louisville: Public investments, private profits, and shared risks. In Gregory D. Squires (Ed.), *Unequal partnerships: The political economy of urban redevelopment in postwar America* (pp. 202-220). New Brunswick, NJ: Rutgers University Press.

Darden, J. T., Hill, R. C., Thomas, J., & Thomas, R. (1987). *Detroit: Race and uneven development.* Philadelphia: Temple University Press.

Dawley, D. (1992). *A nation of lords: The autobiography of the vice lords* (2nd ed.). Chicago: Waveland.

Delaware River Port Authority. (1996). *The Patco Hi-Speed Line: A 25 year retrospective.* Philadelphia: Author.

Department of City Planning. (1993). *Annual report on social indicators.* New York: City of New York.

Department of City Planning. (1995). *Proposed consolidated plan: Federal fiscal year 1996* (Vols. 1-3). New York: City of New York.

Department of City Planning. (1996). *The newest New Yorkers: 1990-1994.* New York: City of New York.

Department of Finance. (1996, July). *Historical county population estimates and components of change, 1990-1996.* Retrieved from the World Wide Web: http://www.dof.ca.gov/html/demograp/rendat.htm

Derthick, Martha. (1972). *New towns in town.* Washington, DC: The Urban Institute.

Douglas, C. C. (1970, June). The curse of contract buying. *Ebony Magazine,* pp. 25, 43-46.

Downs, Anthony. (1981). *Neighborhoods and urban development.* Washington. DC: Brookings Institution.

Downs, Anthony. (1994). *New visions for metropolitan America.* Washington, DC: Brookings Institution.

Downs, Anthony. (1997). The challenge of our declining big cities. *Housing Policy Debate, 8*(2), 359-408.

Dreier, Peter. (1989). Economic growth and economic justice in Boston: Populist housing and jobs policies. In Gregory D. Squires (Ed.), *Unequal partnerships: The political economy of urban redevelopment in postwar America* (pp. 35-58). New Brunswick, NJ: Rutgers University Press.

Dreier, Peter. (1996). Urban politics and progressive housing policy: Ray Flynn and Boston's neighborhood agenda. In W. Dennis Keating, Norman Krumholz, & Phil Star (Eds.), *Revitalizing urban neighborhoods* (pp. 63-82). Lawrence: University of Kansas Press.

Dreier, Peter. (1997). The new politics of housing. *Journal of the American Planning Association, 63*(1), 5-27.

Ducharme, Donna. (1997). *An economic development initiative for North Lawndale: A report to the Steans Family Foundation.* Chicago: Steans Family Foundation.

Dunn, Marvin, & Stepick, Alex, III. (1992). Blacks in Miami. In Guillermo J. Grenier & Alex Stepick III (Eds.), *Miami now! Immigration, ethnicity and social change* (pp. 41-56). Gainesville: University Press of Florida.

Fainstein, Susan S., Fainstein, Norman, Hill, Richard, Judd, Dennis, & Smith, Michael Peter. (1983). *Restructuring the city: The political economy of urban redevelopment.* New York: Longman.

Feagin, Joe R., Gilderbloom, John I., & Rodriguez, Nestor. (1989). The Houston experience: Private-public partnerships. In Gregory D. Squires (Ed.), *Unequal partnerships: The political economy of urban redevelopment in postwar America* (pp. 240-258). New Brunswick, NJ: Rutgers University Press.

Ferman, Barbara. (1996). *Challenging the growth machine: Neighborhood politics in Chicago and Pittsburgh.* Lawrence: University Press of Kansas.

Fine, S. (1989). *Violence in the model city: The Cavanagh administration, race relations, and the Detroit riot of 1967.* Ann Arbor: The University of Michigan Press.

Fisher, Robert. (1994). *Let the people decide: Neighborhood organizing in America* (2nd ed.). Boston: Twayne.

Florida Center for Urban Design and Research. (1988, November). *The Overtown connection—Opportunity for partnership and progress.* Prepared for the St. John Community Development Corporation, Miami, FL.

Florida Center for Urban Design and Research. (1993, December). *Overtown CRP executive summary.* Prepared for Overtown Advisory Board, Inc. and City of Miami, Miami, FL.

Frey, William H. (1995). Immigration and internal migration "flight" from U.S. metropolitan areas: Toward a new demographic Balkanisation. *Urban Studies, 32*(4-5), 733-757.

Frieden, B. J., & Kaplan, M. (1990). Rethinking neighborhood strategies. In N. Carmon (Ed.), *Neighborhood policy and programmes: Past and present* (pp. 238-248). New York: St. Martin's.

Frisbee, M. (1991). *An alley in Chicago: The ministry of a city priest.* Kansas City, MO: Sheed and Ward.

Gale, Dennis E. (1996). *Understanding urban unrest: From Reverend King to Rodney King.* Thousand Oaks, CA: Sage.

Gans, Herbert J. (1962). *The urban villagers.* New York: Free Press.

Gans, H. J. (1995). *The war against the poor.* New York: Basic Books.

Garrow, D. J. (1986). *Bearing the cross: Martin Luther King, Jr. and the Southern Christian Leadership Conference.* New York: Vintage.

Gelfand, Mark I. (1975). *A nation of cities: The federal government and urban America, 1933-1965.* New York: Oxford University Press.

Giloth, Robert. (1996). Social justice and neighborhood revitalization in Chicago: The era of Harold Washington, 1983-1987. In W. Dennis Keating, Norman Krumholz, &

Phil Star (Eds.), *Revitalizing urban neighborhoods* (pp. 83-95). Lawrence: University of Kansas Press.

Giloth, R., Meima, J., & Wright, P. (1984). *Tax, title, and housing court search: Property research for action.* Chicago: Center for Urban Economic Development, University of Illinois at Chicago.

Goetz, E. G. (1993). *Shelter burden: Local politics and progressive housing policy.* Philadelphia: Temple University Press.

Goetz, Edward G. (1996). The community-based housing movement and progressive local politics. In W. Dennis Keating, Norman Krumholz, & Phil Star (Eds.), *Revitalizing urban neighborhoods* (pp. 164-178). Lawrence: University of Kansas Press.

Goldsmith, William W., & Blakely, Edward J. (1992). *Separate societies: Poverty and inequality in U.S. cities.* Philadelphia: Temple University Press.

Gordon, Wayne, L. (1995). *Real hope in Chicago.* Grand Rapids, MI: Zondervan.

Gottlieb, Paul D. (1997). Neighborhood development in the metropolitan economy: A policy review. *Journal of Urban Affairs, 19*(2), 163-182.

Grant, David M., Oliver, Melvin L., & James, Angela D. (1996). African Americans: Social and economic bifurcation. In R. Waldinger & M. Bozorgmehr (Eds.), *Ethnic Los Angeles* (pp. 379-411). New York: Russell Sage Foundation.

Green, Howard Whipple. (1931). *Population characteristics by census tracts, Cleveland, Ohio, 1930.* Cleveland: Plain Dealer Publishing Company.

Greene, Richard. (1991, January). Poverty concentration measures and the urban underclass. *Economic Geography, 67*(1), 240-252.

Greenleigh Associates. (1968). *North Lawndale redevelopment project: Phase 1 report.* New York: Greenleigh Associates.

Greenleigh Associates. (1969). *Lawndale Center: A strategy and action plan* (Prepared with the Lawndale People's Planning and Action Conference). New York: Greenleigh Associates.

Greer, Edward. (1979). *Big steel: Black politics and corporate power in Gary, Indiana.* New York: Monthly Review Press.

Greer, Scott. (1965). *Urban renewal and American cities.* New York: Bobbs-Merrill.

Griego, Linda. (1997). *Rebuilding LA's urban communities: A final report from RLA.* Santa Monica, CA: Milken Institute.

Haar, Charles M. Haar. (1975). *Between the idea and the reality.* Boston: Little, Brown.

Halpern, R. (1995). *Rebuilding the inner city: A history of neighborhood initiatives to address poverty in the United States.* New York: Columbia University Press.

Hartman, Chester. (1974). *Yerba Buena: Land grab and community resistance in San Francisco.* San Francisco: Glide.

Hartman, Chester. (1984). *The transformation of San Francisco.* Totowa, NJ: Rowman & Allenheld.

Hartman, C. (1986). Housing policies under the Reagan administration. In R. G. Bratt, C. Hartman, & A. Meyerson (Eds.), *Critical perspectives on housing* (pp. 362-376). Philadelphia: Temple University Press.

Harvey, D., & Chatterjee, L. (1974). Absolute rent and the structuring of space by governmental and financial institutions. *Antipode, 6*, 22-36.

Hays, R. Allen. (1995). *The federal government and urban housing: Ideology and change in public policy* (2nd ed.). Albany: State University of New York Press.

Hellmuth, Obata, & Kassabaum Sport Facilities Group. (1986, August 1). *Atlanta Stadium Study.* Atlanta: Author.

Hirsch, Arnold R. (1983). *Making the second ghetto: Race and housing In Chicago, 1940-1960.* Cambridge, UK: Cambridge University Press.

Hiskey, Michele. (1991, July 26). Stadium's interest in two lots upsets neighbors. *Atlanta Journal/Constitution*, p. D2.

Homefront. (1977). *Housing abandonment in New York City.* New York: Author.

"Hough City councilwoman Fannie Lewis saw her Hough neighborhood decline—she decided she wasn't going to let that happen without a fight." (1996, September 22). *Plain Dealer* (magazine section), p. 10.

Houghteling, L. (1978). *Pyramidwest: The broad-based approach.* Cambridge, MA: Center for Community Economic Development.

Hughes, Mack A., & Sternberg, Julie. (1992). *The new metropolitan reality: Antipoverty strategy where the rubber meets the road.* Washington, DC: Brookings Institution.

Jackson, Kenneth T. (1985). *Crabgrass frontier.* New York: Oxford University Press.

Jargowsky, Paul A. (1997). *Poverty and place: Ghettos, barrios and the American city.* New York: Russell Sage Foundation.

Jones, B., & Bachelor, L. (1986). *The sustaining hand.* Lawrence: University Press of Kansas.

Jordan, Gertrude. (1989, May 31). HUD warns CMHA of sanctions. *Plain Dealer*, p. 1-B.

Kasarda, John D. (1993). Inner-city concentrated poverty and neighborhood distress: 1970 to 1990. *Housing Policy Debate, 4*(3), 253-302.

Kasarda, John D., & Ting, Kwok-fai. (1996) Joblessness and poverty in America's cities: Causes and policy prescriptions. *Housing Policy Debate, 7*(2), 387-419.

Keating, Larry, Creighton, Max, & Abercrombie, Jon. (1996). Community development: Building on a new foundation," In David L. Sjoquist (Ed.), *The Olympic legacy: Building on what was achieved* (pp. 13-20). Atlanta: Research Atlanta.

Keating, W. Dennis. (1994). *The suburban racial dilemma: Housing and neighborhoods.* Philadelphia: Temple University Press.

Keating, W. Dennis. (1997). The CDC model of urban development, a reply to Randy Stoecker. *Journal of Urban Affairs, 19*(1), 29-33.

Keating, W. Dennis, Krumholz, Norman, & Metzger, John. (1989). Cleveland: Post-populist, public-private partnerships. In Gregory D. Squires (Ed.), *Unequal partnerships: The political economy of urban redevelopment in postwar America* (pp. 121-141). New Brunswick, NJ: Rutgers University Press.

Keating, W. Dennis, Krumholz, Norman, & Star, Phil. (Eds.). (1996). *Revitalizing urban neighborhoods.* Lawrence: University of Kansas Press.

Keating, W. Dennis, & Smith, Janet. (1996). Neighborhoods in transition. In W. Dennis Keating, Norman Krumholz, & Phil Star (Eds.), *Revitalizing urban neighborhoods* (pp. 24-38). Lawrence: University of Kansas Press.

Keyes, Langley C., Schwartz, Alex, Vidal, Avis C., & Bratt, Rachel. (1996). Networks and nonprofits: Opportunities and challenges in an era of federal devolution. *Housing Policy Debate, 7*(2), 201-229.

Kirp, David L., Dwyer, John P., & Rosenthal, Larry A. (1995). *Our town: Race housing and the soul of suburbia.* New Brunswick, NJ: Rutgers University Press.

Kotlowitz, Alex. (1991). *There are no children here: The story of two boys growing up in the other America.* New York: Anchor/Doubleday.

Kozol, Jonathan. (1991). *Savage inequalities: Children in America's schools.* New York: Crown Publishers.

Krumholz, Norman. (1996). The provision of affordable housing in Cleveland: patterns of organizational and financial support. In Willem Van Vliet (Ed.), *Affordable housing and urban redevelopment in the United States* (pp. 52-72). Thousand Oaks, CA: Sage.

Kusmer, Kenneth. (1978). *A ghetto takes shape*. Chicago: University of Illinois Press.

Kusmer, Kenneth. (1995). Black Cleveland in the Central-Woodland Community, 1865-1930. In W. Dennis Keating, Norman Krumholz, & David Perry (Eds.), *Cleveland: A metropolitan reader* (pp. 265-282). Kent, OH: Kent State University Press.

Leete, Laura, with Bania, Neil. (1996). *Assessment of the geographic distribution and skill requirements of jobs in the Cleveland-Akron metropolitan area*. Cleveland: Cleveland Center for Urban Poverty and Social Change, Case Western Reserve University.

Lemann, N. (1991). *The promised land*. New York: Alfred A. Knopf.

Logan, John R., & Molotch, Harvey L. (1987). *Urban fortunes: The political economy of place*. Berkeley: University of California Press.

London, R., & Puntenney, D. (1993). *A profile of Chicago's poverty and related conditions*. Evanston, IL: Center for Urban Affairs and Policy Research, Northwestern University.

Manhattan Borough President. (1995). *Work to be done* (Report of the Borough President's Task Force on Education, Employment and Welfare). New York: Author.

Massey, Douglas S., & Denton, Nancy A. (1993). *American apartheid: Segregation and the making of the underclass*. Cambridge, MA: Harvard University Press.

Mathews, Ethel Mae, & Stuart, Duane. (1990, December 19). Letter to the editor. *Atlanta Journal Constitution*, p. A12.

Mayer, Neil S. (1984). *Neighborhood organizations and community development: Making revitalization work*. Washington, DC: Urban Institute Press.

McPherson, J. A. (1972, April). In my father's house there are many mansions—and I'm going to get me some of them too: The story of the Contract Buyers League. *Atlantic Monthly*, pp. 52-82.

McRoberts, F. (1995, February 1). $51 million phoenix seen in North Lawndale. *Chicago Tribune*, pp. 1-2.

Medoff, Peter, & Sklar, Holly. (1994). *Streets of hope: The fall and rise of an urban neighborhood*. Boston: South End Press.

Mendics, Chris. (1996, June 13). N.J. may punish Camden city on its enterprise zone. *Philadelphia Inquirer*, p. B1.

Metropolitan Housing and Planning Council. (1967). *Model cities in the Greater Chicago area: A 1967 perspective*. Chicago: Metropolitan Housing and Planning Council Papers, University of Illinois at Chicago Library Special Collection.

Mier, Robert. (1993). *Social justice and local development policy*. Newbury Park, CA: Sage.

Mieszkowski, Peter M., & Mills, Edwin S. (1993). The causes of metropolitan suburbanization. *Journal of Economic Perspectives, 7*(3), 135-147.

Mincy, Ronald B., & Weiner, Susan J. (1993). *The underclass in the 1980s: Changing concept, constant reality*. Washington, DC: The Urban Institute.

Mohl, Raymond A. (1983). Miami: The ethnic cauldron. In Richard M. Bernard & Bradley R. Rice (Eds.), *Sunbelt cities: Politics and growth since World War II* (pp. 65-99). Austin: University of Texas Press.

Mohl, Raymond A. (1993). Race and space in Miami. In Arnold R. Hirsch & Raymond A. Mohl (Eds.), *Urban policy in twentieth century America* (pp. 116-143). New Brunswick, NJ: Rutgers University Press.

Mollenkopf, John H. (1983). *The contested city*. Princeton, NJ: Princeton University Press.

Moore, Lee. (1995, April 10). Who's really running Camden? *Courier Post*, pp. B1-B4.

Moynihan, Daniel P. (1969). *Maximum feasible misunderstanding: Community action in the war on poverty.* New York: Free Press.

Nedo, Harold T. (1998, July 2). Camden taken over by control board. *Courier Post*, p. 1A.

Neighborhood Capital Budget Group. (1996). *Completed and planned infrastructure improvements to the North Lawndale community (24th Ward): 1990 through 2001.* Chicago: Neighborhood Capital Budget Group.

Nelson, William, & Meranto, Phillip. (1977). *Electing black mayors: Political action in the black community.* Columbus: Ohio State University Press.

Nelson A. Rockefeller Institute of Government. (1997). *Empowerment zone initiative.* Albany, NY: Author.

Nenno, Mary K. (1989, March-April). Housing and community development after Reagan: A new cycle of policies and partners. *Journal of Housing*, 75-82.

Nightingale, Demetra Smith, & Haveman, Robert H. (Eds.). (1995). *The work alternative: Welfare reform and the realities of the job market.* Lanham, MD: University Press of America.

Norman, Jack. (1989). Congenial Milwaukee: A segregated city. In Gregory D. Squires (Ed.), *Unequal partnerships: The political economy of urban redevelopment in postwar America* (pp. 178-201). New Brunswick, NJ: Rutgers University Press.

North Lawndale Economic Development Corporation. (1974). *Annual report: "It's about time."* Chicago: Author.

North Lawndale Economic Development Corporation. (1975). *Annual report: "Putting the pieces together."* Chicago: Author.

Nyden, Philip, Maly, Michael, & Lukehart, John. (1996). The emergence of stable racially and ethnically diverse urban communities: A case study of nine U.S. cities. *Housing Policy Debate, 8*(2), 491-534.

Olson, Susan. (1973). *Cleveland's urban renewal experience.* Cleveland: Cleveland City Planning Commission.

Olympic Development Program. (1992, July 22). Atlanta: City of Atlanta.

Ong, Paul, & Valenzuela, Abel, Jr. (1996). The labor market: Immigrant effects and racial disparities. In R. Waldinger & M. Bozorgmehr (Eds.), *Ethnic Los Angeles* (pp. 165-191). New York: Russell Sage Foundation.

Orr, M. E., & Stoker, G. (1994, September). Urban regimes and leadership in Detroit. *Urban Affairs Quarterly, 30*, 48-73.

Ott, Dwight, Phillips, Nancy, & Jennings, John Way. (1998, November 13). State takes police role in Camden. *Philadelphia Inquirer*, p. B1.

Palen, J. J. (1987). *The urban world.* New York: McGraw-Hill.

Parachini, L. F., Jr. (1980). *The political history of the Special Impact Program.* Cambridge, MA: Center for Community Economic Development.

Parillo, Rosemary. (1994, January 19). Arnold's crying about being left out. *Courier Post*, p. B1.

Perez, Lisandro. (1992). Cuban Miami. In Guillermo J. Grenier & Alex Stepick III (Eds.), *Miami now! Immigration, ethnicity and social change* (pp. 83-108). Gainesville: University Press of Florida.

Perin, Constance. (1977). *Everything in its place.* Princeton, NJ: Princeton University Press.

Peterson, George E., & Lewis, Carol W. (Eds.). (1986). *Reagan and the cities.* Washington, DC: The Urban Institute.

Pogge, Jean. (1992). Reinvestment in Chicago neighborhoods: A twenty-year struggle. In Gregory D. Squires (Ed.), *From redlining to reinvestment: Community responses to urban disinvestment* (pp. 113-148). Philadelphia: Temple University Press.

Portes, Alejandro, & Bach, Robert L. (1985). *Latin journey: Cuban and Mexican immigrants in the United States*. Berkeley: University of California Press.

Project 80. (1982). *A plan for North Lawndale*. Chicago: Author.

Porter, Michael E. (1995, May-June). The competitive advantage of the inner city. *Harvard Business Review*, 55-71.

Portes, Alejandro, & Stepick, Alex. (1993). *City on the edge: The transformation of Miami*. Berkeley: University of California Press.

Regalado, James A. (1992). Political representation, economic development policymaking, and social crisis in Los Angeles, 1973-1992. In G. Riposa & C. Dersch (Eds.), *City of Angels* (pp. 160-179). Dubuque, IA: Kendall Hunt.

Rich, M. J. (1993). *Federal policymaking: National goals, local choices, and distribution outcomes*. Princeton, NJ: Princeton University Press.

Rich, Wilbur. (1989). *Coleman Young and Detroit politics: From social activist to power broker*. Detroit: Wayne State University Press.

Riordan, Kevin. (1993, September 5). Plan envisions housing, green space and jobs. *Courier Post*, p. 16A.

Riordan, Kevin. (1996a, June 12). Camden to inventory every vacant building. *Courier Post*, pp. B1-B3.

Riordan, Kevin. (1996b, September 11). Chief Pugh to retire. *Courier Post*, pp. A1-A3.

Riordan, Kevin. (1996c, December 17). Sylvia's owner abandons Camden. *Courier Post*, p. B1.

Rosenfeld, R. A., Reese, L. A., Georgeau, V., & Walmsley, S. (1995). Community development block grant spending revisited: Patterns of benefit and program institutionalization. *Publius: The Journal of Federalism*, 25(4), 55-72.

Roughton, Bert, Jr. (1990, November 29). Fight vowed over '96 stadium. *Atlanta Journal/Constitution*, p. A1.

Rouse, Ewart. (1996, July 14). Conflict strikes Camden project. *Philadelphia Inquirer*, pp. B1, B6.

Rusk, David. (1993). *Cities without suburbs*. Washington, DC: Woodrow Wilson Center Press.

Rybczynski, W. (1996). *City life*. New York: Touchstone Books.

St. John's Community Development Corporation, Inc. (1996). *A three year plan for the re-development of Overtown*. Miami: Author.

Saltman, Juliet. (1990). *A fragile movement: The struggle for neighborhood stabilization*. Westport, CT: Greenwood.

Sands, G. (1997, June). *Michigan's renaissance zone program*. Paper presented at the annual meeting of International Sociological Association, Alexandria, VA.

Sbragia, Alberta. (1989). The Pittsburgh model of economic development: Partnership, responsiveness, and indifference. In Gregory D. Squires (Ed.), *Unequal partnerships: The political economy of urban redevelopment in postwar America* (pp. 101-120). New Brunswick, NJ: Rutgers University Press.

Schubert, M. F. (1993). *A housing strategy for North Lawndale: Achieving the dual vision of neighborhood and family stability*. Chicago: The Community Initiatives Program of The John D. and Catherine T. MacArthur Foundation.

Shaw, T. C. (1996, August-September). *We are what we say we are . . .: Race, regime, and representation in Detroit's community development politics, 1985-1993*. Paper presented at the annual meeting of the American Political Science Association, San Francisco.

Sinzinger, Keith. (1986, August 19). Cleveland: The simultaneous rise and fall of an American city. *Washington Post.*

Smith, Jane. (1997). A dialogue on the Atlanta project. In Thomas D. Boston and Catherine L. Ross (Eds.), *The inner city: Urban poverty and economic development in the next century.* New Brunswick, NJ: Transaction Publishers.

Smith, Michael Peter, Guagnano, Gregory A., & Posehn, Cath. (1989). The political economy of growth in Sacramento: Whose city? In Gregory D. Squires (Ed.), *Unequal partnerships: The political economy of urban redevelopment in postwar America* (pp. 260-288). New Brunswick, NJ: Rutgers University Press.

Soja, Edward. (1989). *Postmodern geographics.* London: Verso.

Squires, Gregory. (Ed.). (1989). *Unequal partnerships: The political economy of urban redevelopment in postwar America.* New Brunswick, NJ: Rutgers University Press.

Squires, Gregory D. (Ed.). (1992). *From redlining to reinvestment: Community responses to urban disinvestment.* Philadelphia: Temple University Press.

Squires, G. D., Bennett, L., McCourt, K., & Nyden, P. (1987). *Chicago: Race, class, and the response to urban decline.* Philadelphia: Temple University Press.

Stack, John F., Jr., & Warren, Christopher L. (1992). The reform tradition and ethnic politics: Metropolitan Miami confronts the 1990s. In Guillermo J. Grenier & Alex Stepick III (Eds.), *Miami now! Immigration, ethnicity, and social change* (pp. 160-185). Gainesville: University Press of Florida.

Stephens, Scott. (1997, July 28). Issue pits jobs against kids. *Plain Dealer,* pp. 1-A, 8-A.

Stepick, Alex III. (1992). The refugee nobody wants: Haitians in Miami. In Guillermo J. Grenier & Alex Stepick III (Eds.), *Miami now! Immigration, ethnicity and social change* (pp. 57-82). Gainesville: University Press of Florida.

Stevens, C., & Bivins, L. (1996, December 8). Detroit businesses deliver on road to empowerment. *Detroit Free Press.* Retrieved from the World Wide Web: http://www.detnews.com/empower/index.htm

Stoecker, Randy. (1997). The CDC model of urban redevelopment: A critique and an alternative. *Journal of Urban Affairs, 19*(1), 1-22.

Stone, Clarence N. (1989). *Regime politics: Governing Atlanta, 1946-1988.* Lawrence: University Press of Kansas.

Stone, M. E. (1993). *Shelter poverty: New ideas on housing affordability.* Philadelphia: Temple University Press.

The story of South Los Angeles. (1992, May 11). *Los Angeles Times,* p. A23.

Stowers, Genie N. L., & Vogel, Ronald K. (1994). Racial and ethnic voting patterns in Miami. In George E. Peterson (Ed.), *Big city politics, governance and fiscal constraints* (pp. 63-84). Washington, DC: The Urban Institute Press.

Swanstrom, Todd. (1985). *The crisis of growth politics: Cleveland and Kucinich.* Philadelphia: Temple University Press.

Teepen, Tom. (1990, December 9). Atlanta's Olympic sport: Singing the NIMBY blues. *Atlanta Journal/Constitution,* p. G7.

Thomas, June Manning. (1989). Detroit: The centrifugal city. In Gregory D. Squires (Ed.), *Unequal partnerships: The political economy of urban redevelopment in postwar America* (pp. 142-160). New Brunswick, NJ: Rutgers University Press.

Thomas, J. M. (1995, August). Applying for empowerment zone designation: A tale of woe and triumph. *Economic Development Quarterly, 9*(3), 212-224.

Thomas, J. M. (1997). Model Cities revisited: Issues of Race and Empowerment. In J. M. Thomas & M. Ritzdorf (Eds.), *Urban planning and the African American community: In the shadows* (pp. 143-163). Thousand Oaks, CA: Sage.

Turner, Margery A., Struyk, Raymond J., & Yinger, John. (1991). *Extract from Housing Discrimination Study: Synthesis.* Washington, DC: The Urban Institute.

Unger, Irwin. (1996). *The best of Intentions.* New York: Doubleday.

Urban Land Institute. (1986). *North Lawndale: An evaluation of redevelopment potential and strategies for the North Lawndale community of Chicago.* Washington, DC: Author.

U.S. Department of Commerce. (1993). *1990 census of population and housing, file STF3A, Dade County and city of Miami, Florida.* Washington, DC: Bureau of the Census.

U.S. Department of Housing and Urban Development. (1995). *Empowerment: A new covenant with America's communities.* Washington, DC: HUD.

U.S. Department of Labor. (1994). *State and metropolitan area employment and unemployment, Miami and Dade County, Florida.* Washington, DC: Bureau of Labor Statistics.

U.S. General Accounting Office. (1985, September 30). *Federal funds promised, provided, and used in Dade County, Florida, after the May 1980 civil disturbances* (GAS/HRD-85-88). Washington, DC: Author.

Vergara, Camilo Jose. (1995). *The new American ghetto.* New Brunswick, NJ: Rutgers University Press.

Vickers, Robert J. (1996, November 24). Federal funds "empowering" few residents in grant zone. *Plain Dealer,* pp. 1-A, 16-A.

Vidal, Avis. (1992). *Rebuilding communities: A national study of urban community development corporations.* New York: Community Development Research Center, New School for Social Research.

Vidal, Avis. (1996). CDCs as agents of neighborhood change: The state of the art. In W. Dennis Keating, Norman Krumholz, & Phil Star (Eds.), *Revitalizing urban neighborhoods* (pp. 149-163). Lawrence: University of Kansas Press.

Vidal, Avis C. (1997). Can community development re-invent itself? The challenges of strengthening neighborhoods in the 21st century. *Journal of the American Planning Association, 63*(4), 429-438.

Walker, Paulette J. (1990, December 27). Stadium area links survival to Olympics. *Atlanta Journal/Constitution,* p. E1.

Walsh, Joan. (1997). *Stories of community building and the future of urban America.* New York: Rockefeller Foundation.

Warren, Christopher L., Corbett, John G., & Stack, John F., Jr. (1990). Hispanic ascendancy and tripartite politics in Miami. In Rufus P. Browning, Dale Rogers Marshall, & David H. Tabb (Eds.), *Racial politics in American cities* (pp. 155-178). New York: Longman.

What's wrong with Camden's schools (seven-part series). (1996, December 8-13). *Courier Post.*

Wikstrom, G., Jr. (1974). *Municipal response to urban riots.* San Francisco: R & E Research Associates.

Williams, T., & Bakama, B. (1992). *The role of the church in community development: Two case studies.* Minneapolis: Hubert H. Humphrey Institute of Public Affairs.

Wilson, C. A. (1992). Restructuring and the growth of concentrated poverty in Detroit. *Urban Affairs Quarterly, 28*(2), 187-205.

Wilson, James Q. Wilson (Ed.). (1966). *Urban renewal: The record and the controversy.* Cambridge, MA: MIT Press.

Wilson, William Julius. (1987). *The truly disadvantaged: The inner city, the underclass, and public policy.* Chicago: University of Chicago Press.

Wilson, William J. (1996). *When work disappears.* New York: Alfred A. Knopf.

Witzke, Dorward R. (1967). *Wrap-up report.* Cleveland: Cleveland Little Hoover Commission.

Wolff, R. C. (1990). *Sister Henrietta of Hough.* Chicago: Loyola University Press.

Wolman, Jerry and Associates. (1965). *City within a city: Proposal for a new town on the Camden waterfront.* Philadelphia: Author.

Wood, R. (1990). Model cities: What went wrong—the program or its critics? In N. Carmon (Ed.), *Neighborhood policy and programmes: Past and present* (pp. 61-73). New York: St. Martin's.

Wurster, Catherine Bauer. (1959). The dreary deadlock of public housing. *Architectural Forum, 106*(5), 140-142, 219, 221.

Wylie, J. (1989). *Poletown: Community betrayed.* Urbana: University of Illinois Press.

Zeller, George. (1993, October). *Job loss in Cleveland and Cuyahoga County.* Cleveland: Council of Economic Opportunity.

Zielenbach, S. (1998). *The art of revitalization: Improving conditions in inner city neighborhoods.* Unpublished doctoral dissertation, Northwestern University.

Zimmerman, Emily, & Li, Jiali. (1995). *Low-income populations in New York City: Economic trends and social welfare programs, 1994.* New York: United Way of New York City and City of New York, Human Resources Administration.

# INDEX

217

# ABOUT THE AUTHORS

**Thomas Angotti** is Associate Professor in and Chair of the Graduate Center for Planning and Environment at Pratt Institute in Brooklyn, New York. He is the author of *Metropolis 2000: Planning Poverty and Politics* (1993), *Housing in Italy* (1977), and many articles in professional journals. He has worked and written extensively on urban planning and community development in the United States, Latin America, and Europe. He is a Fellow at the American Academy in Rome, Associate Editor for America of *Planning Practice and Research*, Participating Editor for *Latin American Perspectives*, and Editor of *Planners Network*. He was previously a city planner with the New York City Department of City Planning and has worked for the state governments in New Jersey and Massachusetts. He taught at the graduate level at SUNY, Columbia University, Harvard, and the University of California, Berkeley. He holds a Ph.D. in Urban Planning and Policy Development from Rutgers University.

**Robert A. Catlin** has worked as a professional urban planner with local governments and as a planning consultant since 1961 in Minneapolis, Southern California, Baltimore, Washington, D.C., New York City, and

Florida. His teaching and research interests include planning history and theory, housing and community development, urban revitalization, neighborhood planning, land use policy, and race as a factor in planning and public policy formulation. His books include *Racial Politics and Urban Planning: Gary, Indiana* (1993) and *Land Use Planning, Environmental Protection, and Growth Management: The Florida Experience* (1997). He holds a Ph.D. from Claremont Graduate School (1976) and an M.S. from Columbia University in urban and regional planning (1972).

**Mittie O. Chandler** is Associate Professor and Director of the Master of Urban Planning, Design, and Development Program and the Master of Science in Urban Studies Program housed in the Levin College of Urban Affairs at Cleveland State University. Her work experience includes positions as a city planner and a public housing manager in Detroit. She has been active with numerous community-based organizations in Detroit and Cleveland. She is currently conducting an assessment of the Cleveland Supplemental Empowerment Zone. She holds a master's degree in urban planning and Ph.D. in political science from Wayne State University.

**Dennis E. Gale** is Department Chair and Henry D. Epstein Professor of Urban and Regional Planning, Florida Atlantic University. He is the author of *Understanding Urban Unrest: From Reverend King to Rodney King* (Sage, 1996), *Washington, D.C.: Inner City Revitalization and Minority Suburbanization* (1987), and *Neighborhood Revitalization and the Postindustrial City: A Multinational Perspective* (1984). Formerly, he was Professor and Director of the Center for Washington Area Studies at George Washington University and a research director at The Urban Institute in Washington, D.C.

**Robert Giloth** is a Senior Associate at the Annie E. Casey Foundation in Baltimore, Maryland. He manages investments in workforce development and community economic development. He recently edited the book *Jobs and Economic Development: Strategies and Practices*. Formerly, he directed CDCs in Chicago and Baltimore, and served as Deputy Commissioner of Economic Development in the mayoral administration of Harold Washington. He holds a Ph.D. in city and regional planning (1988) from Cornell University.

**Larry Keating** is Associate Professor in the Graduate Program in City Planning at the Georgia Institute of Technology, where he teaches planning theory and history, as well as housing and community development practicums. He is a cofounder of the Community Design Center of Atlanta (CDCA) and currently serves as Treasurer and frequent collaborator. CDCA conducts applied research on issues affecting low-income people and provides technical assistance in planning, architecture, and community and real estate development to low-income neighborhoods. He also codirects the Georgia Tech–Georgia State University–CDCA–Community Outreach Partnership Center. He has worked with the Peoplestown neighborhood since 1990. He holds a doctorate in urban and regional planning from the University of Wisconsin–Madison.

**W. Dennis Keating** is Professor of Law and Urban Affairs, Chair of the Department of Urban Studies, and Associate Dean of the Levin College of Urban Affairs at Cleveland State University. He teaches and researches housing and community development, neighborhood planning, and land use law. He has authored numerous articles and book chapters. His latest edited book is *Revitalizing Urban Neighborhoods* (1996), and he is the author of *Rent Control: Regulation and the Rental Housing Market* (1998). He has participated in policy evaluations, which presently include the Cleveland's Empowerment Zone program and neighborhood planning. His Ph.D. in city and regional planning is from the University of California, Berkeley, and his J.D. is from the University of Pennsylvania.

**Norman Krumholz** is Professor in the Levin College of Urban Affairs at Cleveland State University. Previously, he served as an urban planning practitioner in Buffalo, Pittsburgh, and Cleveland. He was Cleveland's city planning director from 1969 to 1979 and was a member of President Jimmy Carter's National Commission on Neighborhoods. He is a past president of the American Planning Association (1987) and recipient of the APA's Distinguished Leadership Award (1990). In 1991, his book *Making Equity Planning Work* (with John Forester) won the Paul Davidoff Award of the Association of Collegiate Schools of Planning. He is the author of numerous books, articles, and book chapters on urban planning practice, theory, and neighborhood development.

**Ali Modarres** is Professor in the Department of Geography and Urban Analysis, California State University, Los Angeles. He is also the director of the Applied Research Program at the Edmund G. "Pat" Brown Institute of Public Affairs on this campus. He specializes in urban geography, and his primary research and publication interests are immigration, race, and ethnicity in American cities. He has published a number of articles on environmental issues, especially urban transportation and demand management policies, within the last two decades. He holds a master's degree in landscape architecture and a doctorate from the University of Arizona.

**Kenneth Reardon** is Associate Professor in Urban and Regional Planning at the University of Illinois at Urbana-Champaign, where he engages in research, teaching, and service activities focused on the empowerment efforts of low-income urban communities and serves as a faculty coordinator for the University's East St. Louis Action Research Project. He serves as the Co-Chairperson of the Planners Network, a national association of progressive civic leaders and planning professionals, committed to promoting social equality, racial justice, and a healthy environment.

R

R